3D Printing with Autodesk 123D®, Tinkercad®, and MakerBot®

Lydia Sloan Cline

New York   Chicago   San Francisco   Athens   London   Madrid
Mexico City   Milan   New Delhi   Singapore   Sydney   Toronto

Cataloging-in-Publication Data is on file with the Library of Congress

McGraw-Hill Education books are available at special quantity discounts to use as premiums and sales promotions, or for use in corporate training programs. To contact a representative, please visit the Contact Us pages at www.mhprofessional.com.

3D Printing with Autodesk 123D®, Tinkercad®, and Makerbot®

1 2 3 4 5 6 7 8 9 0   QVS/QVS   1 0 9 8 7 6 5 4

ISBN 978-0-07-183347-9
MHID   0-07-183347-1

Sponsoring Editor Roger Stewart	**Indexer** Karin Arrigoni
Editorial Supervisor Jody McKenzie	**Production Supervisor** Jean Bodeaux
Project Manager Howie Severson, Fortuitous Publishing	**Composition** Fortuitous Publishing
Copy Editor Bart Reed	**Art Director, Cover** Jeff Weeks
Proofreader Richard Camp	**Cover Designer** Tom Lau

To Makers everywhere.

And a big thanks to...

Roger, Amber, and Christie, who indulge my mad scientist projects.

Guillermo Melantoni and Christian Pramuk, program managers of the Autodesk 123D suite, for their assistance with this book.

The Autodesk company, whose products enable Makers all over the globe to draw their dreams and earn their livings.

About the Author

Lydia Cline has been using Autodesk products in her work for
architecture firms since AutoCAD version 1.0 was released. She judges
competitive technology events, writes textbooks, creates open source
educational materials, and teaches at Johnson County Community
College in Overland Park, Kansas.

Contents

Preface

THE PAST 10 YEARS have seen an explosion of interest and potential in 3D printing. Novices and professionals are using it to make their lives and jobs more rewarding and productive. When this technology is combined with websites such as Quirky, Kickstarter, Indiegogo, Prosper, Etsy, Shapeways, Ponoko, and Facebook, people are empowered to turn hobbies into businesses, start side businesses, and leave jobs they no longer like or that have left them. 3D printing is truly a part of the New Industrial Revolution.

Thought leaders compare current 3D printers to the dot-matrix printers and the Apple Macintosh of the early 1980s. Being competitive will eventually require knowledge of this technology because it will surely disrupt and innovate in diverse fields. Electronics, toy, food, and automotive companies are experimenting with it, as is the construction industry and the military. 3D printing solutions are being considered as answers to substandard housing in developing countries and in other applications we can barely imagine right now.

Creating, manufacturing, and advertising is becoming democratized due to affordable tools and low start-up costs. No matter what your age or background is, you can participate; the barriers are low. The same way that software and laser printers let anyone publish from their kitchen table, and WordPress and websites let anyone open a nice storefront, now anyone with knowledge of 3D software and printing technology can put out a product. Although no one foresees an end to mass manufacturing, it's a good bet that micro-manufacturing and just-in-time manufacturing catering to niche and customization markets will get bigger (see Figure 1). One size no longer fits all.

Websites such as Thingiverse and Instructables, where designs are freely shared, let the world tinker with, customize, and improve each other's work in the spirit of open source. Maker Faires let them show off what they've done.

The MakerBot company has crowd-sourced an effort to get a 3D printer in every classroom. Making may (and should) eventually be part of the curriculum. The nascent Maker Movement might counter cheap overseas labor and bring manufacturing back to the U.S. 3D printing is for teachers, students, homeschoolers, dreamers, entrepreneurs, and the merely curious—and because you're holding this book, you're obviously interested in it! You may already have some concrete ideas about what you'd like to do. So, congratulations in finding your way here, because now you'll learn how to turn those ideas into a physical product.

Let's get started.

Figure 1 At www.makie.me, customers can select features for dolls, which are then 3D-printed and shipped to them. These dolls were displayed at Autodesk University.

Resources to Check Out

- **Autodesk University 2013 video "The New Industrial Revolution"**   http://au.autodesk.com/au-online/classes-on-demand/class-catalog/2013/class-detail/3604.

- **Information about Maker Faires**   www.makerfaire.com.

- **Kickstarter, Indiegogo, and Prosper**   Sites where entrepreneurs can solicit crowd-funding, market research, and feedback from a large participant community. www.kickstarter.com, www.indiegogo.com, and www.prosper.com.

- **Quirky**   A large community that gives feedback on the product development and manufacturing process. www.quirky.com.

- **Etsy**   An online marketplace for handmade products. www.etsy.com.

- **Shapeways, Ponoko, and Imaterialise**   Online service bureaus where you can send files for 3D printing or CNC cutting, and sell copies via their storefronts. www.shapeways.com, www.ponoko.com, and http://i.materialise.com/.

- **Thingiverse**   Download all kinds of things to 3D print. www.thingiverse.com.

- **Instructables**   Get instructions to make all kinds of things. www.instructables.com.

- Anderson, Chris. Makers, *The New Industrial Revolution*, Crown Business: New York, 2012.

- Lipson, Hod. *Fabricated: The new world of 3D printing*, Wiley: Hoboken, New Jersey, 2013.

CHAPTER 1

Hello, Maker!

WHAT IS A *MAKER*? A maker is a person who creates things. Thinks up things. Fixes and hacks things. Is always looking for "the next best thing," or ways to improve existing things. A maker is a producer, an artist, an inventor. The Maker Movement is a worldwide movement, with millions of people starting small businesses, selling their products, or just decorating their lives with tangible outputs of their ideas. Makers congregate at hacker spaces to create and collaborate, and at Maker Faire events to show, sell, and build relationships (Figure 1-1). The fact that you're reading this book indicates that *you're* a maker, or are interested in becoming one. Fabulous, because it's a great time to be a creative or entrepreneurial person!

Figure 1-1   The Kansas City Maker Faire

What Is Making?

Making is basically Do-It-Yourself-ing. Historically, DIYers have done their thing by sewing, cooking, crafting, woodworking, building, machining, programming, repurposing, and tinkering. Today, modeling programs combined with 3D printers let DIYers take their game to a whole new level. They can design, iterate, customize and prototype their products themselves relatively easily and cheaply. Popular 3D-printed products include shoes, jewelry, eyeglasses, phone accessories, doll accessories, toys, promotional trinkets, and replacement parts—all can be made with plastics, metals, sterling, and ceramics.

At a higher level, makers are designing and printing prosthetics, food, buildings, guns, rocket parts, and even living body organs. The designs are done with software called a *modeling program*. Physical production is done with fabrication tools such as 3D printers and CNC machines. Figure 1-2 shows a printed architectural model, and Figures 1-3 through 1-6 show four whimsical prints from Shapeways .com, a website where makers display and sell their creations via print on demand (POD).

Figure 1-2   A floor plan that was digitally modeled and printed

Figure 1-3   A Spirograph-inspired coaster modeled with 3ds Max (courtesy Luke Sozzo)

Figure 1-4   A metal rabbit whistle modeled with Sculptris (courtesy www .pookas.de)

Figure 1-5   A bowtie modeled with Rhinoceros (courtesy www .monocircus.com)

Figure 1-6   A Christmas ornament modeled with MODO (courtesy Rob Parthoens, www .baroba.be)

What, Exactly, Is a Modeling Program?

A modeling program is graphics software in which you create three-dimensional (3D) drawings called models. You can spin a drawing around on the screen to view it from any angle and generate two-dimensional (2D) drawings from it, which is useful for producing construction or manufacturing documents. Modeling is different from traditional computer-aided drafting (CAD) software, which is basically an electronic pencil with which you draw 2D pictures. With modeling, your picture is always 3D.

Although many ready-made models are available for download from websites such as the 123D Gallery, Thingiverse, the Trimble 3D Warehouse, Turbosquid, GrabCad, Formfonts, and the Makerbot Digital Store, knowing how to operate modeling software enables you to create your own designs. Maybe you have better ideas, such as a great accessory for the Google Glass or some custom Lego blocks. Furthermore, many ready-made designs require tweaking to print well, or maybe you want to customize them. This is done inside modeling software.

Many modeling programs exist to choose from. The Autodesk company produces the industry-standard AutoCAD, Revit, Inventor, 3ds Max, and Maya. Programs from other sources include Rhinoceros, SolidWorks, SketchUp, OpenSCAD, Blender, Sculptris, and Modo. However, they all have steep learning curves and some have steeper prices. Autodesk 123D suite will get you up and running quickly in the modeling and 3D printing world. It will also give you foundational skills for the more advanced programs.

Mesh vs. Solid Models

The type of model a software program produces affects how you work with it. There are two types: mesh and solid.

A mesh model is a hollow form made of polygons (flat surfaces). Think air-filled balloon. Figure 1-7 shows two mesh models. Note that although their forms appear circular, they are actually multiple polygons.

A solid model has a continuous surface and volume like a rock. 123D Design is a solid

modeler; Figure 1-8 shows two models made with it. The sphere and cylinder are true circular forms and both are completely filled inside. Their volumes define their forms.

Mesh modelers have more flexible tools than solid modelers and are typically used for organic, freeform subjects. They enable you to model things that would be difficult or impossible with a solid modeler, such as animated characters, people, and animals (see Figure 1-9). That said, mesh models are prone to forming holes

Figure 1-8 A solid model is a continuous mass.

Figure 1-9 This dragon would be impossible to do with a solid modeling program. (Source: thingiverse.com.)

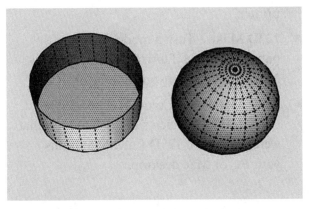

Figure 1-7 A mesh model is made of multiple polygons and is hollow inside.

during the modeling process that make them unprintable. These must be healed, which is another process.

Solid modelers function by adding, subtracting, and intersecting forms, and are typically used for making mechanical parts. A solid model often takes less preparation to 3D print because it's more likely to be "watertight"—meaning if you poured water into it, none would leak out. No gaps.

Proficiency with both model types is desirable because you can combine them for the best of both worlds. For example, you could make a mesh model of your pet's head (impossible to do with a solid modeler) and stick it on a coffee mug (which is a snap to make with a solid modeler).

So Tell Me about the Autodesk 123D Suite Already!

Autodesk 123D is a suite of simple, powerful modeling apps that are integrated with premade content, social media, and printing/fabrication services. If you're a casual creator, it's for you. The suite is constantly evolving, but as of this writing, the apps and their platforms are as follows:

- **123D Catch**   Take multiple photos of a subject with a digital camera, iPad, or iPhone or Android device, and upload them to Autodesk's servers, where they're returned as a photorealistic mesh model covered with photos of the subject (called *phototexturing*) (see Figure 1-10). Then send this to a 3D printer or CNC machine cutter.

- **123D Circuits**   Design an electronic project online in the standard "breadboard" view, make professional printed circuit boards with built-in tools, and then send the boards off for manufacturing.

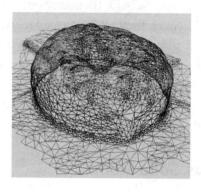

Figure 1-10   A phototextured model made with an iPad and 123D Catch. The mesh is under the photos.)

- **123D Creature**   Design character mesh models on your iPhone or iPad from scratch and then 3D print them.

- **123D Design**   Create a solid model on your PC or Mac desktop, online, or on your iPhone, iPad, or iPod. Then send it to a 3D printer.

- **123D Make**   Turn a model into a cutting pattern for CNC fabrication on your PC or Mac, online, or on your iPhone or iPad.

- **123D Meshmixer** Combine mesh and solid models or fix problems in mesh models that were made with 123D Catch. Do it this on your PC or Mac desktop.

- **123D Sculpt**   Modify a ready-made character mesh model online with an iPhone or iPad. Then share it online or send to an Apple AirPrint printer.

- **The CNC Utility**   Generate G-code for a model, a language that enables a CNC machine to physically make it.

- **The Sandbox**   An online lab where you can peek in on apps being developed and tested.

- **Tinkercad**   Like 123D Design, this is a Web-based solid modeler, but targeted to the youngest makers. While simpler than 123D Design, it remains robust, with a unique feature called "shape generators." Many use it to create custom objects to import into the *Minecraft* open-world video game.

The 123D apps roughly parallel Autodesk's professional software. For instance, 123D Design is similar to Autodesk Fusion 360, a Web-based version of Inventor, which is a professional-level solid modeling program. 123D Catch is similar to Autodesk Recap (reality capture), a program that makes models from photographs of a subject. Catch and 123D Creature can be imported into Autodesk 3ds Max, Maya, or Mudbox for further development. 123D Design files can be exported as *.dwg* files (a Premium feature) and then used in AutoCAD or any other *dwg*-compatible program.

Be aware that the 123D suite is relatively new (it debuted in 2011) and its developers frequently iterate it. Therefore, by the time you read this book, some features may not work exactly as described, and others may have been added, changed, or removed. The website itself may also look different. Stratasys, the company that owns the Makerbot 3D printer, is constantly developing and rolling out new models, too. Consider it all part of the challenge of being an early adopter.

What Is a 3D Printer?

A 3D printer is a machine that builds a physical model from a digital model. It does this by adding successive layers of material onto a flat surface called a *build plate* and melts that material into the shape of the model. Commercial and consumer-grade printers are available in a variety of sizes and types based on their intended product. For instance, there are printers that can make chocolate bars, novelty foods, ceramics, and gold jewelry.

There are almost as many 3D printer choices as there are modeling software choices. The Makerbot (Figure 1-11) was chosen for this book due to its popularity in the "prosumer" market and its ability to directly receive 123D app files. It is currently priced from about $1,700 to $6,500, depending on the model. The plastic filament it uses is sold in 0.2 kilogram (0.5 pound), 1 kilogram (2.2 pounds), 5 pound, and 10 pound spools. There's an optional Makercare service plan. You can also buy a Makerbot, Makercare, and a Premium 123D subscription all bundled together for a special price.

Buy a Makerbot from its website or from one of its retail stores (currently three, in New York City, Boston, and Greenwich, Connecticut). You

Figure 1-11   A Makerbot Replicator 2 and three rolls of filament.

can also buy some models from Autodesk and Amazon (the latter of which will save you the shipping charge). All methods typically require a few weeks to fulfill the order.

What Is a CNC Machine?

A Computer Numerical Control (CNC) machine is one in which a computer controls the fabrication process (Figure 1-12). CNC cutters include routers, lathes, and lasers, and they work by removing material to shape the item. They cut flat shapes from cardboard, plastic, metal and wood (Figure 1-13).

Consumer-level CNC machines are available; the ShopBot company makes popular models for home use, but they, like the Makerbot, are still too expensive for many makers. Luckily, these tools are found in makerspaces.

What Is a Makerspace?

A *makerspace*, also called a *hackerspace*, is a physical location containing computers with software, hardware, and fabrication tools for member or public use (Figure 1-14). Some libraries and museums have them. TechShops are set up in some states; a $125/month membership fee provides you access to a workshop with at least a million dollars' worth of equipment. Then there's 100kGarages.com, a matchmaker service that connects people who own shop equipment with people who want to use it. Google "makerspace" or "hackerspace" and the name of your city to find what's available near you.

So What's This Book About?

This book is an introduction to digital and physical modeling. No prior knowledge of any drafting or modeling programs is assumed or needed. If you're a beginner, you've come to the

Figure 1-12   A CNC laser cutter

Figure 1-13   A violin made with a CNC router. The files for the violin were made with Corel Draw.

Figure 1-14   Hammerspace, a makerspace in Kansas City, MO.

right place! You'll learn six apps from the 123D suite—Design, Catch, Meshmixer, Tinkercad, Make, and the CNC Utililty. Most of the chapters are written to stand alone so that you can go straight to the one you're interested in instead of reading them all chronologically. And when a new tool is introduced, it's discussed separately from the tutorial project so you can look it up if you just want to quickly read how it works.

We'll make solid models with Design and generate construction drawings. We'll turn physical objects into phototextured mesh models with Catch, and fix them up in Meshmixer. We'll also use Meshmixer to alter models made in Design and prepare them for printing. We'll make a cutting pattern for a CNC machine with Make, and generate G-code for it with the CNC Utility. We'll learn how to use Tinkercad and check out its shape generators. Finally, we'll set up a Makerbot and print one of our creations with it. If you don't have your own Makerbot, there are online service bureaus that print models for a price, and we'll look at those, too. Along the way, we'll grab stuff from the 123D Gallery and Thingiverse. All projects are done in a step-by-step, tutorial manner, but if you like videos, check out my YouTube channel, where videos of some of this book's projects, plus other projects, are posted. Its URL is listed at the end of this chapter.

Autodesk 123D's Free and Pay Accounts

You'll need to open an Autodesk account to download the 123D apps and utilize the Web-based ones. This account is free, and the user ID and password you choose will log you in to all Autodesk websites. Go to https://accounts .autodesk.com/ to create an account. Any payments you make, such as for a premium subscription or premium content, are managed

through it. Files you create are saved to this account, and notifications from Autodesk arrive through it via the Messages feature (the Messages feature is not for contacting other users).

The 123D suite is free to use, but has pay options that offer more features. All account and product prices in this book are current at the time of writing, but like the software itself, are subject to change. Here's what the free 123D account offers:

- You can use the 123D apps to create models for noncommercial purposes.
- You can download models from the 123D Gallery.
- You can download 10 Premium models a month created by the 123D master modelers.
- You have unlimited cloud space in which to save your models. Save them privately or post them in the gallery for the world to admire and download.

Here's what the $9.99 monthly Premium membership offers:

- You can use the 123D apps to create models for commercial purposes.
- You can download unlimited Premium models created by the 123D master modelers.
- You can access the LayOut feature. This converts a 123D Design file into a *.dwg* file, which you can open inside Autodesk 360, a free Web app, to generate and annotate orthographic views. You can also import the *.dwg* file into AutoCAD or any *.dwg*-compatible software to further develop it.
- You get a Pro account on Instructables.com, a popular how-to website.
- You get unlimited private circuit designs for the Circuits app and a 5 percent discount on printed circuit board orders.

Here's what a $99 one-year or $189 two-year membership offers:

- You get one free 3D print per year, delivered to your home (single color, maximum size 4"×4"×4"). Choose one from the gallery or upload one of your own.

- You get a $40 or $90 discount code to apply to a Makerbot purchase.

Here's what a free Tinkercad account offers:

- Access to the basic editor

- Unlimited storage for your designs

- The ability to import 3D mesh

- The ability to import 2D drawings

- The ability to use shape generators (JavaScript programs that construct 3D shapes).

- A personal $19-per-month Tinkercad subscription enables you to use your designs commercially and manage licenses (that is, you can determine who can use them and how).

- To cancel a subscription or to get help with purchased products, submit a support ticket at http://sitesupport.123dapp.com/home.

What You Need, Computer-Wise

Here's the minimum needed to run the 123D suite:

PC

- Windows 7 (32-bit or 64-bit), Windows XP Professional or Home Edition (SP3), or Windows XP Professional x64 Edition (SP2)

- Intel Core 2 Duo or later, AMD Athlon 64 X2 or later processor

- At least 2GB RAM

- At least 200MB of free disk space

- OpenGL graphics card with 256MB or more dedicated graphics memory (the stronger, the better)

- 1280×1024 minimum screen resolution

- Internet connection

Mac

- Apple Mac OS X, version 10.7 or higher

- Mac computer with an Intel Core 2 Duo, Core i3, Core i5, Core i7, or Xeon processor

- At least 2GB RAM

- 1280×1024 minimum screen resolution

- Internet connection (preferably fast, because the online platform often does not load well with a slow connection)

Tinkercad is a Web app that runs on Windows (Vista or newer), Apple (OS X 10.6 or newer), or Linux in any browser that supports HTML5/WebGL. If you get a message stating that your browser doesn't support WebGL when you try to run it, update the browser. Google Chrome 10 or newer works the best on all the Web apps even when another browser is listed as supported.

A two-button mouse with a scroll wheel is very helpful. The scroll wheel performs zoom and pan functions, saving you from having to click those icons all the time.

File Formats

The 123D suite imports and exports various file formats (also called "extensions"). A *format* is the group of characters to the right of the period in a file's name that describes how the file's information is encoded. Some formats are proprietary, meaning they can only be used in one software program, one company's programs, or one printer. Others are open standard, meaning they can be used in many different programs and printers. I'll be tossing out file format names throughout this book, so instead of describing them in each chapter, I've listed them here for you to reference as needed.

- **.123dx**   This is the 123D Design desktop file format.

- **.3dmk**   This is the 123D Make file format.

- **.3dp**   This is a proprietary Autodesk reality capture model format.

- **.3dp**   This is a common format for reality capture models. 123D Catch uses it.

- **.amf**   This an alternative to the .stl format. It contains color and texture information.

- **.avi**   This is an audio/video file that can be watched on Apple QuickTime or Windows Media Player.

- **.dwf**   This is an Autodesk Web format file that compresses 2D and 3D CAD files in a manner that preserves properties, sheets, and layers. The file can then be e-mailed or posted online, and recipients can view and edit it with free Autodesk programs.

- **.dwg**   This is a proprietary AutoCAD drawing file format. Premium members can export a 123D Design model into a .dwg file.

- **.dwx**   This is a locked AutoCAD drawing file.

- **.dxf**   This is an exchange format that allows 2D AutoCAD files to be opened and worked on in many other CAD programs.

- **.fbx**   This is an exchange format that allows models to be opened and worked on in multiple software programs. It contains color and texture information. It is an Autodesk format, but can be exported from some non-Autodesk programs.

- **.obj**   This is another format used by many 3D printers, and is created within a modeling program. When combined with *.mtl* and *.jpg* files, which the modeling program creates at the same time, color and texture information can be incorporated into a full-color 3D print or imported into another software program.

- **.ply**   This is an alternative to the *obj* format. It contains color, texture, and transparency information without needing an accompanying .mtl or .jpg file. It also supports different properties for the front and back of a polygon.

- **.sat, .smb, .step, .stp**   These are exchange formats that allow files to be opened and worked on in multiple software packages.

- **.stl**   This format is used by many 3D printers and is created within a modeling program. Both solid models and mesh models can be exported as .stl files, but the .stl file itself is a mesh file. It does not contain color or texture information; therefore, the 3D print will simply be the color and texture of the printer's filament.

- **.svg**   This is an open standard vector graphics format for artwork made in illustration programs.

- **.thing**   This is a proprietary Makerware (software that runs the MakerBot) format. You can export a file in this format, reimport it into Makerware later, and continue to edit it.

- **.x3d**   This is a format used by open source 3D printing software. It contains color and texture information, and is the successor to the .vrml file format.

- **.xrml**   This is a publishing language that can be attached to a file to provide digital rights management.

Download 123D Design Desktop

Before we leave this chapter, let's download Design Desktop. Point your browser to http://www.123dapp.com and go get it! Click the Apps menu at the top of the page and select 123D Design from the submenu (see Figure 1-15). This will take you to links for a PC or Mac download.

Figure 1-15   Access the download links at www.123dapp.com.

project managers, and fans of the software often hang out there. Post your questions, and answer other people's questions when you can. Share your tips to help build up this community!

So, ready to get started? Head over to Chapter 2, where you'll start learning how to use the Design app.

Sites to Check Out

- **Lydia's YouTube channel**   www.youtube.com/user/ProfDrafting

- **123D blog**   http://blog.123Dapp.com/

- **Stratasys blog**   http://blog.stratasys.com/

- **List of Autodesk blogs**   http://usa.autodesk.com/adsk/servlet/index?id=4805213&siteID=123112

- **Description of 123D paid account**   www.123dapp.com/gopremium

- **Description of Tinkercad subscription plans**   https://tinkercad.com/plans/

- *Make* **magazine's website**   http://makezine.com/

- **GE's traveling garage website, where makers can view and use its prototyping and manufacturing tools**   www.ge.com/garages

- **TechShop's website**   www.techshop.ws

- **Service that matches people with files and equipment.**   www.100kgarages.com

- **Repositories of models, both free and for a cost:**

  - www.123dapp.com/Search/content/all

  - www.thingiverse.com

  - www.turbosquid.com

  - https://3dwarehouse.sketchup.com

  - www.formfonts.com

  - http://grabcad.com/library

  - https://digitalstore.makerbot.com/

Follow the instructions to install. When you download a future version, it will replace the current one, because two different versions cannot reside simultaneously on your computer. Be aware that if you use an older version of an app, you won't be able to open files that were made with newer versions. When you download Design, you'll be asked if you want to download Meshmixer. This is because Meshmixer optimizes all 123D app files for printing and can send them directly to the printer or service bureau.

One more thing. Check out the support forum at forum.123dapp.com. The suite's developers,

CHAPTER 2

Getting Started with 123D Design

THIS CHAPTER TAKES YOU on an in-depth tour of the Design desktop interface. We'll look at its menus and features to see what they do, and you'll learn the capabilities of this app in the process.

Open 123D Design Desktop

Click the 123D Design desktop icon, shown here, and a splash screen of tips will initially appear. Arrow through them, if interested, and then click the X in the upper-right corner to close this screen. You can check the box in the lower-left corner if you don't want the screen to appear again. (You can always bring the screen back later if you need to reference it.)

Probably the best tip at this stage, however, is to press the F1 key, which brings up a screen of *hot keys* (or shortcut keys), as shown Figure 2-1. You might want to print this screen and keep it handy as you go through the interface. Also, you should know that the RETURN key finishes an operation, the ESC key cancels an operation, and the DELETE key serves as an eraser.

Once the splash screen is closed, the interface is exposed, as shown in Figure 2-2. It consists of

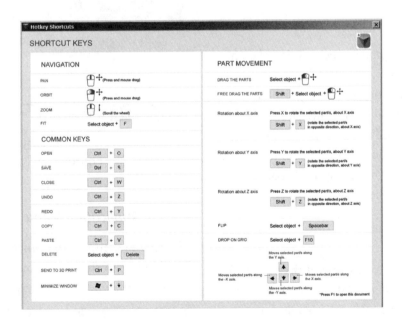

Figure 2-1 A screen of shortcut keys and navigation tips. (A change unacknowledged yet in this screenshot is that the "Drop on Grid" function, which snaps an object to the grid, is now Select + D).

Menu bar

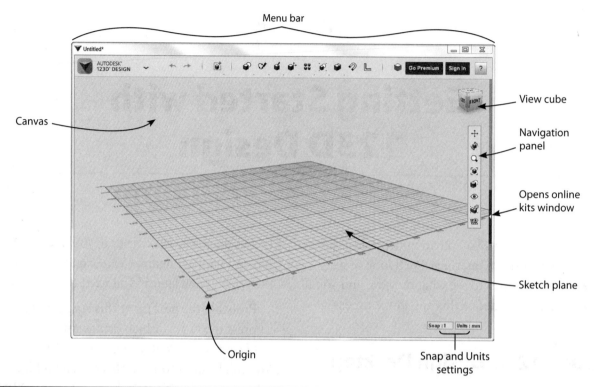

Canvas

View cube

Navigation panel

Opens online kits window

Sketch plane

Origin

Snap and Units settings

Figure 2-2   The 123D Design interface

a sketch plane, menu bar, ViewCube, navigation panel, online kits window, and Snap and Units settings.

Sketch Plane

Also called the *work plane* or *canvas*, the sketch plane is a gridded surface upon which you draw all lines and forms. It can be oriented horizontally or vertically, as shown in Figure 2-3. Sketches and solids snap to the grid lines, making it easy to model in those increments. The origin (point 0,0,0) is in the lower-left corner of the grid, denoted by the large circle.

Menu Bar

The menu bar is a horizontal bar at the top of the screen that contains menus (also called *buckets*) and functions. If you can't see them all, drag the screen window so that it's larger. Let's start with the menu on the far left, as shown in Figure 2-4.

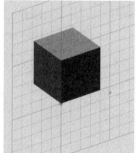

Figure 2-3   The sketch plane is a gridded work surface upon which all lines are drawn. It can be oriented vertically or horizontally.

Design Application Menu

This menu contains the functions described next. You'll need to sign in to your Autodesk account to access the ones that take you to the Web.

- **New**   Opens a new, local file and closes the currently open one (you'll be prompted to save it first, though). If you want multiple

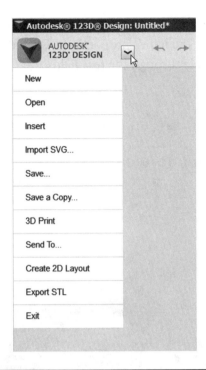

Figure 2-4 The 123D Design Application menu contains utility functions.

Note that there is no option to send the file to a traditional printer. If you want paper copies of your model, you need to take a screenshot. On a PC, press PRINT SCREEN, open the Paint utility, and press CTRL-V to paste the image into it. On a Mac, press COMMAND-SHIFT-4, which temporarily replaces the cursor with a selection window. Click and drag the tool to make your selection. The screenshot will appear on the desktop.

files open at the same time, click the desktop icon to open new files.

- **Open**   Brings up links that let you browse files on your desktop, in your 123D account, or in the 123D Gallery, and to view examples. Before a new file opens, you'll be prompted to save and close the currently opened one. You need to be signed in to your Autodesk account for this option.

- **Insert**   Takes you online to links that let you browse files stored on your desktop, in the cloud, and in the 123D Gallery. It brings the selected file into the currently open one. You need to be signed in to your Autodesk account for this option.

- **Import SVG**   Brings in a local .svg file, with options to make it a sketch or solid.

- **Save**   Preserves all work done on the current file. You can save it locally with the To My Computer option or online with the To My Projects option.

- **Save a Copy**   Saves the open file locally and makes that copy current.

- **3D Print**   Sends the file to Meshmixer for printing preparation and options.

- **Send To**   Sends the file to 123D Make (desktop), the CNC utility, or a 3D print Web service. All these options are discussed in later chapters.

- **Create 2D Layout**   Takes you online to start the process of generating construction (2D) views of the model.

- **Export STL**   Makes a local .stl file of the model.

- **Exit**   Closes the program.

Undo/Redo

This function undoes operations one at a time, all the way to opening the file; it also redoes all undone actions (see Figure 2-5).

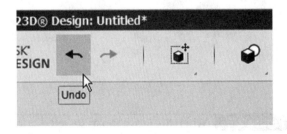

Figure 2-5 Undo and redo all operations, one at a time, by clicking these arrows.

Transform

The Transform menu contains the Move and Scale tools. Move brings up a "box gizmo" that has manipulators for changing the item's linear and planar positions as well as for rotating it (see Figure 2-6). Scale brings up an arrow to drag for size adjustment and a dialog box to enter numbers (see Figure 2-7). Click the drop-down

arrow for the Non Uniform option if you want to change the item's proportions. Click the TAB key to cycle through the text fields.

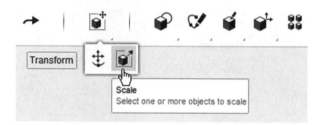

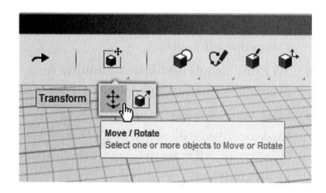

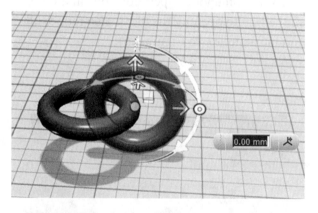

Figure 2-6   When the Move tool is selected, a "box gizmo" appears that lets you to change an item's location and position.

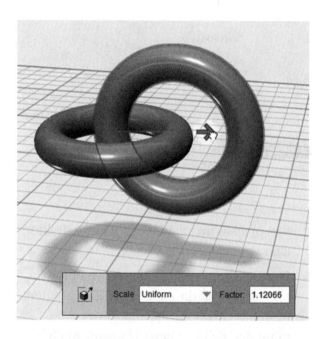

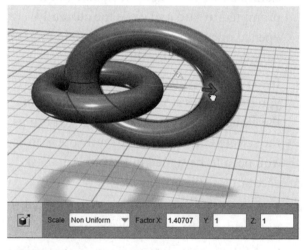

Figure 2-7   The Scale tool lets you to change an item's size, either uniformly or non-uniformly.

Primitives

The Primitives menu contains 3D forms, also called *solids*, and 2D shapes (see Figure 2-8). Click a solid and move (don't drag) it to the sketch plane. This is called *cruising*. When a primitive is first "cruised" from its submenu, it can be snapped to the center of another primitive's face (snapping occurs at the white cruise widget—that is, the "handle" where the mouse grabs it).

When a primitive is cruised into the canvas, a dialog box appears at the bottom of the screen. You can accept the default measurements or type in new ones; press the TAB key to switch text fields. Click the primitive in place after typing the sizes. Primitives cannot be cruised again, nor can their sizes be changed via this dialog box again (instead, use the Scale tool's Non Uniform option, shown in Figure 2-7).

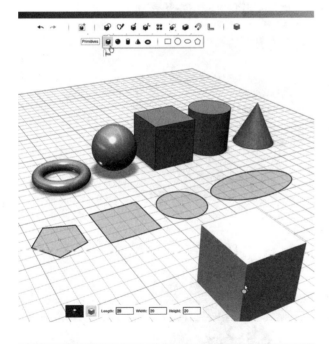

Figure 2-8   The Primitives menu contains simple forms and shapes. They can be "cruised" from the submenu to the sketch plane or to the center of another primitive's face.

Sketch

Some of this menu's tools—Rectangle, Circle, Ellipse, and Polygon—are premade shapes, whereas others, such as Polyline, Spline, 2-Point and 3-Point Arcs, let you draw your own shapes (see Figure 2-9, upper left). Fillet, Trim, and Extend let you edit sketches; Offset copies and places a sketch a specified distance from the original; Project puts a sketch onto another surface. Note the pop-up boxes in Figure 2-9. These are prompts that instruct what steps to take while using a tool. All tools generate them. Pay attention to these boxes, because if an operation isn't working, it may be that you aren't doing what you're prompted to do.

Construct

The Construct menu's tools turn sketches into complex forms by extruding, sweeping, revolving, and lofting them (see Figure 2-10). Extrude turns a sketch into a solid. It also cuts through a solid. Sweep extrudes a sketch along a path. Revolve rotates a sketch around an axis, and Loft interpolates a form between sketches. Each tool has multiple options that are accessed via a drop-down menu on the glyph that appears when activated.

Modify

The Modify menu's tools—Press/Pull, Tweak, Split Face, Fillet, Chamfer, Split Solid, and Shell—change existing solids. Press/Pull lengthens or shortens a solid's face as well as cuts through a solid, and Tweak warps a solid (see Figure 2-11). Chamfer angles an edge, Fillet rounds an edge, and Shell hollows out a solid (see Figure 2-12). Finally, Split Face breaks a face into two faces, and Split Solid breaks a

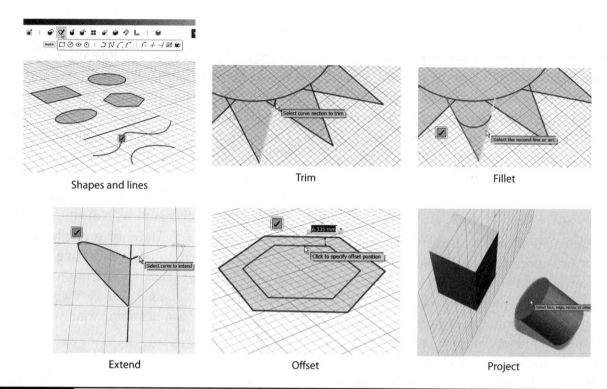

Shapes and lines   Trim   Fillet

Extend   Offset   Project

Figure 2-9 Sketch and edit shapes, polylines, splines, and arcs. Sketch tools can trim, fillet, extend, offset, and project lines.

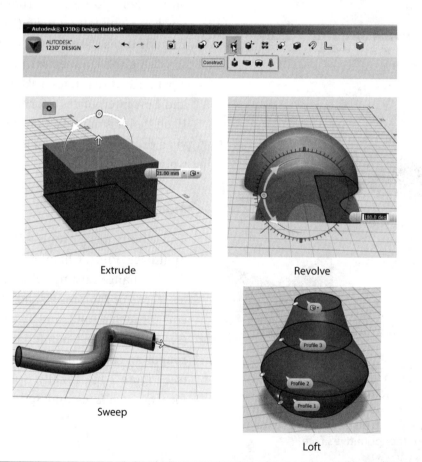

Extrude   Revolve

Sweep

Loft

Figure 2-10 The Construct menu's Extrude, Sweep, Revolve and Loft tools create complex 3D forms from 2D sketches.

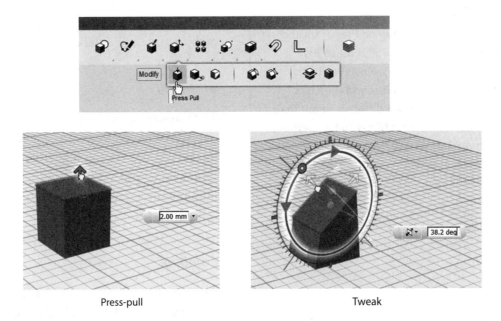

Press-pull

Tweak

Figure 2-11 The Modify menu has tools for changing solids. Shown here are Press/Pull, which lengthens or shortens the face of a solid, and Tweak, which warps it.

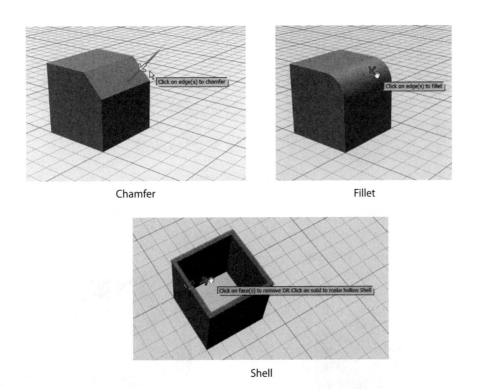

Chamfer

Fillet

Shell

Figure 2-12 Chamfer and Fillet change the model's edges. Shell hollows out a solid.

> What's the difference between Press/Pull and Extrude? In theory, Press/Pull is for solids, and Extrude is for sketches. However, both work similarly on most solids and sketches. Sometimes one will give a different result than the other, but this is hard to predict. Experiment with both on the same surface as your models increase in complexity.

solid into two pieces (see Figure 2-13). Like the Construct tools, each Modify tool has multiple options.

Pattern

Tools on the Pattern menu array (that is, copy and arrange) sketches and solids (see Figure 2-14).

The array can be rectangular, circular, or along a path. A sketch or solid can also be mirrored.

Grouping

The Grouping menu "paperclips" multiple items together so they can be selected and moved as one. They can later be ungrouped. Double-clicking a group highlights the items individually, enabling them to be worked on separately (see Figure 2-15).

Combine

The Combine menu contains the Boolean operations Join, Cut, and Intersect. Unlike grouping, combining alters solids permanently; hence, they cannot later be uncombined. This action can only be done on solids, not shapes or

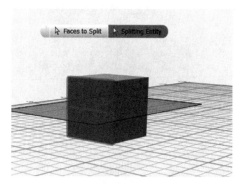
Split face

Result of split face

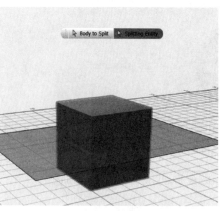
Split solid

Result of split solid

Figure 2-13 Split Face divides a face into two; Split Solid divides the whole form into two forms.

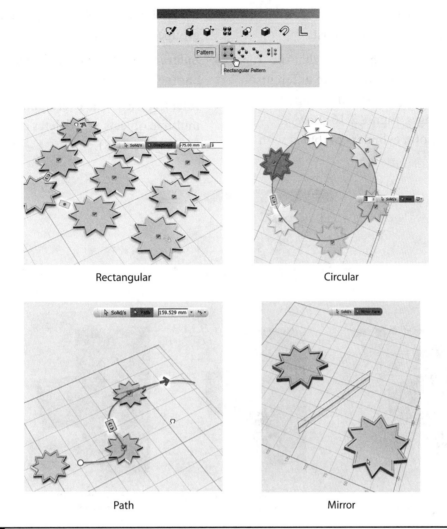

Rectangular

Circular

Path

Mirror

Figure 2-14 Sketches and solids can be arrayed and mirrored.

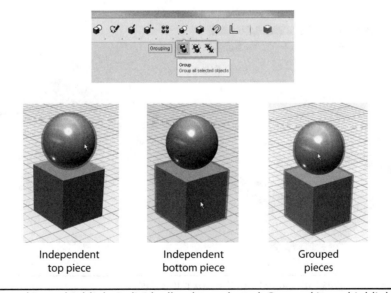

Independent
top piece

Independent
bottom piece

Grouped
pieces

Figure 2-15 Ungrouped items highlight individually when selected. Grouped items highlight together.

shelled (hollowed out) solids. The three Combine actions are detailed in the following list:

- **Merge** welds forms together. Note that the Chamfer and Fillet operations give different results on merged edges than on grouped or single-item edges. Figure 2-16 shows a chamfered edge of a cylinder combined with a box, and a chamfered edge of a cylinder grouped with a box.

- **Subtract** removes one intersecting form from another (see Figure 2-17).

- **Intersect** leaves the overlapping parts of multiple objects and deletes everything else (see Figure 2-18).

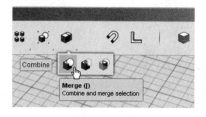

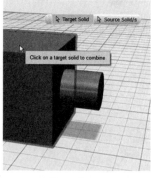 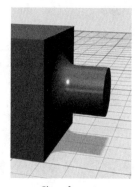 

Two pieces merged   Chamfer on a merged edge   Chamfer on an unmerged edge

Figure 2-16   A chamfer looks different when applied to a combined edge versus a grouped edge.

Figure 2-17   Subtract removes this shape from the box it intersects.

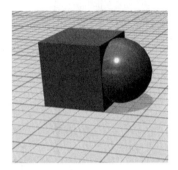 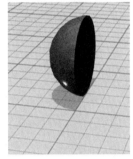

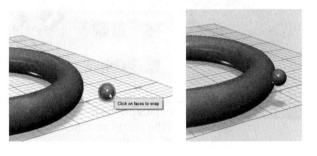

Figure 2-18   Intersect keeps the overlapping parts of multiple solids and deletes everything else.

Figure 2-19   Click on the faces you want to snap together.

Snap

You use the Snap tool to place two items together by clicking their faces (see Figure 2-19).

Measure

The Measure tool tells you the angle, area, volume, and edge length as well as the distance between pieces (see Figure 2-20).

Material

The Material menu offers colors and textures that can be applied to the model (see Figure 2-21). It is only for display purposes only and has nothing to do with a physical 3D print.

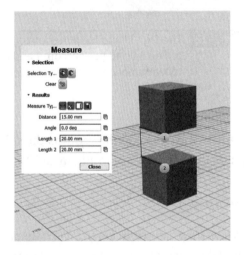

Figure 2-20   Click the edges of these boxes with the Measure tool to obtain their lengths and the distance between them.

Go Premium, Sign In, and Help

The Go Premium menu takes you to a screen with information about Premium (pay) membership benefits and a link to sign up. The Sign In menu, shown in Figure 2-22, takes you to the 123D website, where you can access data about your account, a private message box,

cloud-saved files, and (for Premium members) the LayOut feature. There's also a link to a gallery of models made by other users. The Help menu links to online sources such as the 123D blog, support forum, and tutorial videos.

Figure 2-21 Paint materials on the model by selecting the pieces and then clicking a Materials swatch.

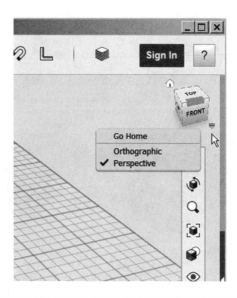

Figure 2-23 The ViewCube shows the model's orientation on the work plane. Mouse over it to make a house icon and drop-down arrow appear.

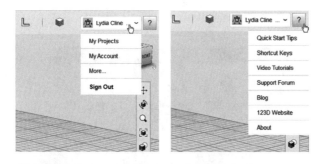

Figure 2-22 The Sign In and Help menus contain online links to your account and places to learn more about the software.

The Quick Start Tips option reopens the splash screen of tips (also shown in Figure 2-22).

ViewCube

The ViewCube, shown in Figure 2-23, displays the model's orientation on the work plane. Click the left mouse button on the Cube, drag to rotate it, and watch the model rotate along with the Cube. Click the Cube's sides to view the model orthographically (that is, as a top, front, or side view). Hover the mouse over the Cube to make a house icon and a drop-down arrow appear. Clicking the house icon returns the sketch plane to its default position; clicking the drop-down arrow provides options to view the model in perspective or orthographically. (The

Go Home option does the same thing as the house icon.) An orthographic view lacks the line convergence of perspective, which is useful when you're aligning forms because you can see their alignments better than in perspective view. A model set to display orthographically will remain that way until changed back.

Navigation Panel

The navigation panel, shown in Figure 2-24, contains tools that let you view and display the model in different ways:

- **Pan** This tool slides the model around the screen. You can pan more efficiently by pressing and holding the mouse's scroll wheel down.

- **Orbit** This tool whirls you, the viewer, around the model, so you can see it from all angles and heights. Your position relative to the model moves, not the model itself. You can orbit more efficiently by dragging the mouse while holding its right button down.

Figure 2-24 Use Pan, Orbit, Zoom, and Fit on the navigation panel to view the model from different angles.

- **Zoom** This tool gives you a magnified view of the model (think telephoto lens) to see small details, or reduces your view of it (think wide-angle lens) so you can see the big picture. You can zoom more efficiently by rotating the scroll wheel up and down.

- **Fit** This tool fills the screen up with the model. If your model hides off in a corner after you click it, you've got little pieces you drew earlier still hanging around. Find and erase them, and your model will come back. Clicking Fit is a handy way to locate "lost" pieces.

If you have a two-button mouse with a middle scroll wheel, roll the wheel up and down with your index finger to zoom in and out; press and hold it down to pan.

Now look at the last four icons on the panel. The box/circle icon displays the model with materials and outlines, materials only, and outlines only (see Figure 2-25). The box/eye icon hides sketches, solid models, and mesh models separately (mesh models being any .stl or .obj files you import). The group/magnet icon offers the option of automatically grouping forms when they're snapped together. By default, this

  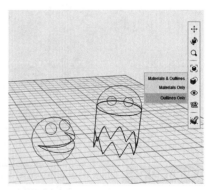

Figure 2-25 Display the model's materials and outlines, materials only, or outlines only.

feature is on (that is, it groups). The grid/eye icon displays or hides the sketch plane.

Online Kits Window

The small white arrow in the blue rectangle (under and to the right of the navigation panel) accesses the online parts bin. Bins are themed collections of fully editable, scaled models made by the 123D team. An Internet connection is needed because these bins are stored online and load after you open the window. Open the models by dragging them onto the canvas. Click the drop-down arrow at the top of the window to see a collapsed menu of other choices. Text, robots, space vehicles, gadgets, and more are available (see Figure 2-26).

You don't have to be signed in to your Autodesk account to access the bins.

Snap and Units

The two functions at the bottom of the screen—Snap and Units—help you model accurately (see Figure 2-27). Snap restricts movement to specific intervals. Select one between 0.1 and 10, or turn Snap off to simply glide the forms and sketches across the work plane. Units lets you choose millimeters (mm), centimeters (cm), or inches (in) to model in. Millimeters are the most precise for 3D printing, but if you're more comfortable working in inches, do so; just change the units to mm before sending the model to the Makerbot.

A file remembers any interface changes you make (for example, selecting new units) after it is closed and reopened. However, new files will still have the default settings, so the changes will need to be reset.

Figure 2-26   Click the white arrow to open the online kits window and then click the drop-down arrow at the top to see a collapsed menu of themed collections. Shown here are the Smart Primitives and Robot Heads kits.

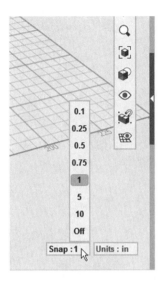

Figure 2-27   Snap and Units offer choices for applying measurements to the model.

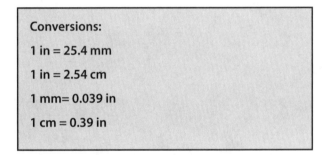

Conversions:

1 in = 25.4 mm

1 in = 2.54 cm

1 mm= 0.039 in

1 cm = 0.39 in

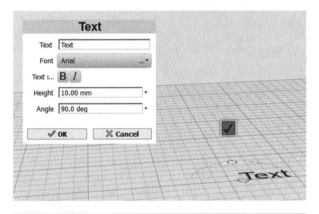

Text Tool

In Chapter 1, I said that the 123D apps quickly iterate. Well, Design just iterated! A Text tool was added to the menu bar (Figure 2-28). Click on the T and then click on the workplane to place it, type in the dialog box that appears, and push ENTER.

Figure 2-28   The Sketch Text tool.

Summary

In this chapter you began learning how to use the 123D Design desktop app by exploring its interface. Now that you know the features, what they do, and where to find them, join me in Chapter 3, where we'll start modeling.

Model It!
123D Design Projects

IN CHAPTER 2, you learned what 123D Design Desktop can do and how to access its features. In this chapter we'll work on nine projects: a studded bangle bracelet, sun sculpture wall decoration, star cookie cutter, Jell-O mold tray, peace sign necklace pendant, personalized business card, Pac-Man and Pinky figures, and a tea cup with a frog inside. All will utilize different tools and techniques. In the process you'll learn how to sketch, use construction and modifying tools, combine solids, import files and access your online account.

Be aware that all models are not 3D-printable. For example, a printable model needs a minimum thickness and cannot have sharp edges. Since this chapter is about learning Design, by itself a large task, printability isn't addressed. Characteristics of a printable model, and how to make an

How to Select and Erase

Before we begin, here's how to perform the essential tasks of selecting and erasing. To select (highlight) a face, click it and then run the mouse over one of the face's edges. A selected face turns dark blue. Click directly on an edge to select it; it will turn black.

To select multiple items, drag your mouse (press the left button and hold) from the upper-left to the lower-right corner. This creates a selection window; everything completely inside this window will be highlighted. Drag the mouse from the lower-right to the upper-left corner to create a *crossing* window; everything the window touches will be highlighted. Alternatively, select one item,

hold the SHIFT key down, and click other items to include them in the selection. Erase something by selecting it and pressing the DELETE key.

When you select an entire object (not just a portion of it), a long glyph appears at the bottom of the screen, as shown below. This glyph provides quick access to tools that move, scale, hide, and add material textures to the selection. From here you can also export the selection into .123dx or .stl formats, send it to Meshmixer's modify and print areas, and send it to Make. Note that you can apply all of these tools to the entire model or to just a selection.

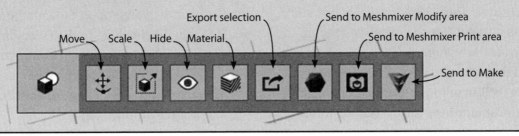

Move   Scale   Hide   Material   Export selection   Send to Meshmixer Modify area   Send to Meshmixer Print area   Send to Make

Figure 3-1   Studded bangle bracelet

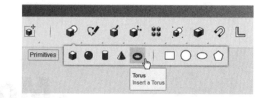

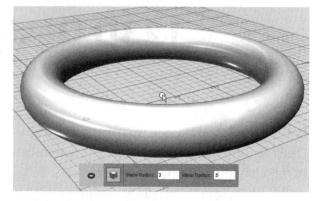

Figure 3-2   Bring a torus onto the canvas, type its dimensions, and click it onto the sketch plane.

unprintable model printable, are discussed in Chapter 9.

Studded Bangle Bracelet Project

To make the studded bangle bracelet shown in Figure 3-1 we'll use tools from the Primitives, Pattern, and Material menus as well as the Move and Snap tools. For this project, set the units to inches. Here are the steps to follow:

1. Make the bracelet body shown in Figure 3-2 by clicking the Primitives menu and cruising a torus onto the canvas. Note the dialog box that appears at the bottom of the screen. Before clicking the torus into place, type **3** for the major (outer) radius, press the TAB key to go to the second field, type **.5** for the minor (inner) radius, and then press ENTER.

2. To add the stud shown in Figure 3-3, click the Primitives menu and cruise a sphere onto the canvas. Type **.25** in the dialog box and press ENTER. Click the sphere to select it, then activate the Snap tool. Click the sphere first and then click the torus. The sphere (first item) will snap to the torus (second item) at the location where you clicked the torus.

> The dialog box is the easiest way to specify a size, and only shows up when the primitive is first cruised onto the canvas. If you want to change an item's size later, use the Scale tool (it has a Non Uniform option that lets you change the x, y, and z axes separately), or you can use the Edit Dimension option discussed later in this chapter.

If you click the torus or sphere, you'll see they now highlight together. That's because snapping automatically *groups* items, or sticks them together (see Figure 3-4). This enables you to select and move them as one. You can still select each piece individually to edit, but some operations are impossible to perform on groups. So with the solids still highlighted, go to the Grouping menu, choose Ungroup, and click either the torus or sphere. They'll become separate again. You can also click Ungroup first and then select the grouped items.

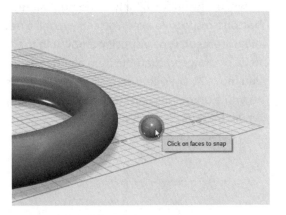

1. Click the Snap tool, then click on the sphere.

2. Click on the torus. The sphere will snap to the torus.

Figure 3-3   Size a sphere and then snap it to the torus.

Figure 3-4   Grouped items highlight as one. Group and ungroup them via the Grouping menu.

Figure 3-5   Turn off automatic snap/grouping by clicking the group/magnet icon in the navigation panel.

If you don't want solids to automatically group upon snapping, click the group/magnet icon in the navigation panel to turn this feature off (see Figure 3-5). We'll keep it off for the rest of the chapter.

3. To move the stud into the bracelet, click the stud to select it. A glyph of tools appears; click Move, which makes the manipulator appear (discussed in the sidebar on the next page). Select the stud, click the Move tool, grab an arrow on the manipulator, and move the stud halfway into the torus.

Move, Rotate, and Copy with the Manipulator

The *manipulator* is a feature that can move, copy, and rotate anything you draw. It's accessed by clicking the Move icon on the tools glyph. It has three arrows, three planes, and three buttons. Drag the arrows to move a selected item up and down, back and forth, or left and right along the arrows' respective axes. Drag the planes to

move an item freely up and down, back and forth, or left and right along the planes' respective axes. Drag the buttons to rotate an item up and down, back and forth, or left and right along the buttons' respective axes. You can eyeball a new measurement, type a specific amount in the text field, or drag and snap the item along the grid as you move it. Dragging snaps it every 5°; pressing ALT while dragging snaps it every 15°; pressing SHIFT while dragging snaps it every 45°.

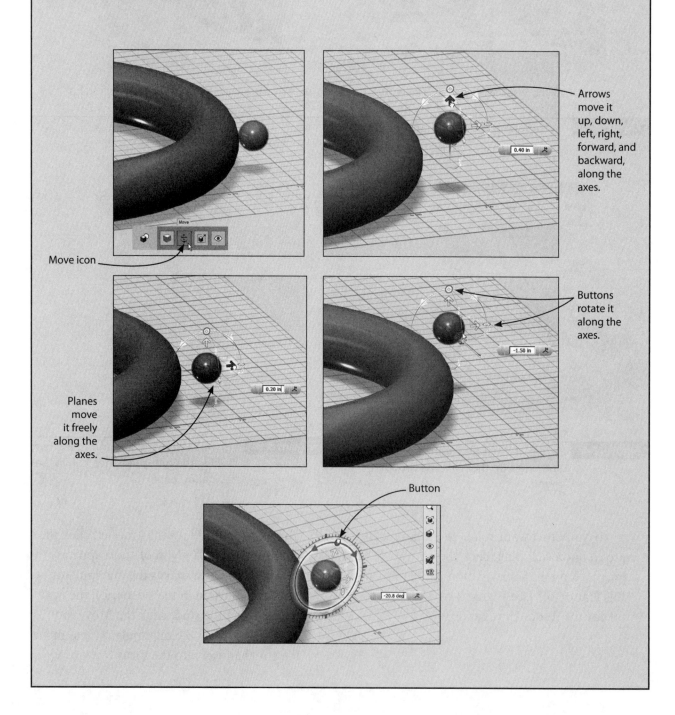

Move icon

Arrows move it up, down, left, right, forward, and backward, along the axes.

Planes move it freely along the axes.

Buttons rotate it along the axes.

Button

To copy an item, select it and press CTRL-C and then CTRL-V. A copy will be made over the original item and the manipulator for it will appear. Move the copy where you want. Repeat this process for each new copy.

The manipulator can also reorient the axis along which a selected item is moved. Click the axis symbol and then click the item to be reoriented.

Drag a button to the angle desired and then click the axis symbol to finish. You can now move the item along the newly aligned axes. This change affects only the selected item and isn't permanent; if you later want to move the item along that axis again, you'll have to reset the axis.

Axis graphic

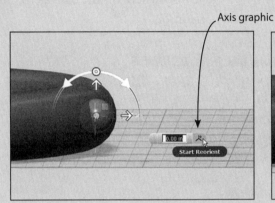

1. Click the axis graphic.

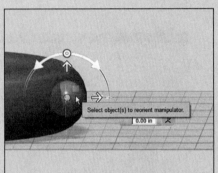

2. Select the object.

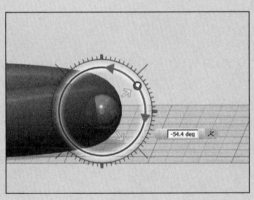

3. Drag the button to the angle wanted.

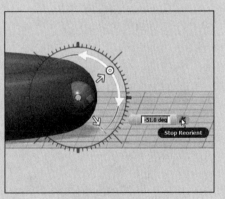

4. Click the icon again.

5. Move the object along the reoriented axis.

4. Array the stud by selecting it, clicking Pattern, and choosing Circular Pattern (see Figure 3-6). A glyph with two buttons—Solid and Axis—appears. Solid is the item we want to array (the stud) and it is already selected. Therefore, click Axis. Next, click anywhere on the bracelet; this will set the torus's center as the axis. Note the drop-down arrow on the glyph; it conceals options. Make sure Full is chosen so that the stud will array completely around the bracelet.

Two arrow buttons will appear that you can drag forward or backward to add or subtract copies. Alternatively, type the number of copies in the glyph's text field. If these buttons don't appear after you click Axis, click the Solid button and then click the stud again. Each copy has a check mark; if you click one, you'll remove that copy. When you have the number of studs wanted, click anywhere on the canvas to finish the operation.

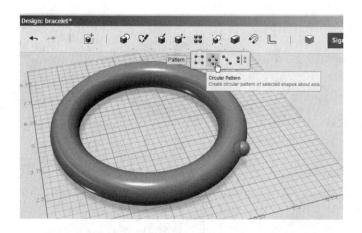

1. Select the stud and choose Circular Pattern.

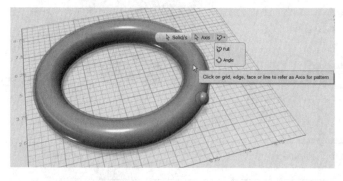

2. Click Axis and select the torus.

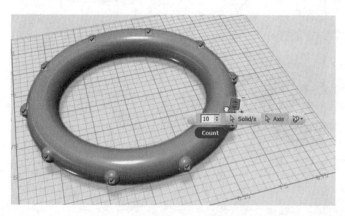

3. Drag the yellow buttons to make copies or type a number in the glyph.

Figure 3-6   Array the stud around the bracelet with the Circular Pattern tool.

5. To add texture and color, click the Material menu (see Figure 3-7). A window of material swatches appears; click a swatch and then click it onto the torus. Then select a stud and click a different material onto it. The material paints whatever surface is selected, so make sure what you don't want to repaint isn't highlighted. You can remove a material with the default swatch, located in the top row.

6. To overlay a color onto any material, first click the color wheel, click the interior diamond, and then click the Apply Overlay box (see Figure 3-8). The color will immediately appear on the model. Colors don't need to overlay a material; you can apply a color directly to the model by clicking the wheel, clicking the interior diamond, and then clicking the model. Uncheck the Apply Overlay box to remove a color.

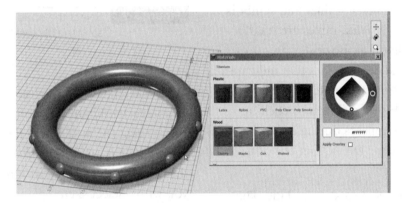

Figure 3-7 Apply color and texture from the Material menu.

Figure 3-8 Overlay color onto material or apply it on its own.

7. Finally, save your work, as shown in Figure 3-9. Click the 123D menu, choose Save, and select To My Computer (it will highlight blue when selected). This saves the file locally. The To My Projects option saves it online, and we'll talk about that later in this chapter.

Our next two projects include sketching. Because it is the foundation of most solid modeling, let's discuss it first.

Sketching: Tools, Planes, and Text

Straight lines are sketched with the Polyline tool, curvy lines with the Spline tool, arcs with the Two Point and Three Point Arc tools, and shapes with their respective shape tools. While you sketch, different *inference* lines and symbols appear (see Figure 3-10). Examples are small geometric shapes that indicate center points or perpendicularity, and dashed lines that indicate a line under construction that matches the length of an existing line. Finish a sketch by clicking the check mark that appears during a sketch operation or by pressing ENTER. Get out of a sketch by pressing ESC.

All sketching is done on a sketch plane. That plane can be oriented horizontally, vertically, or at an angle. Cruise a box primitive onto the canvas before sketching and then click its top, front, or side face to activate a sketch plane in that direction. Clicking an angled face produces an angled sketch plane. Sketching can't be done on curved surfaces.

Sketches are made with three clicks. The first selects the sketch plane, the second selects its starting point, and the third selects its endpoint or size. All lines on a sketch must be on the same plane to be editable and operated on as a whole. If your sketches are not on the same plane, many construction tools will not work on them. Because it's not visually obvious whether

Figure 3-9   Select the option To My Computer to save the file locally.

or not sketches are on the same plane, it's easy to inadvertently put them on different planes. Ensure that the next sketch line gets put on the same plane as an existing sketch line by placing the next sketch's first click—the one that prompts you with "Select sketch plane or click to define the plane"—somewhere on the first sketch (for example, on the face, outline, or endpoint). Don't place it randomly on the grid itself. Place the second click where you want the sketch to start.

You may want to type specific measurements in the pop-up text fields and then click. Alternatively, you may be more concerned about proportions, and just click lines where they look good. You might prefer counting grid squares and utilizing the Snap feature (remember, you can adjust the Snap setting). You can always scale a sketch to the correct size needed after it's finished. You can also edit its size as you work. Edit a line by clicking it; edit a closed face by clicking its outline (not its face).

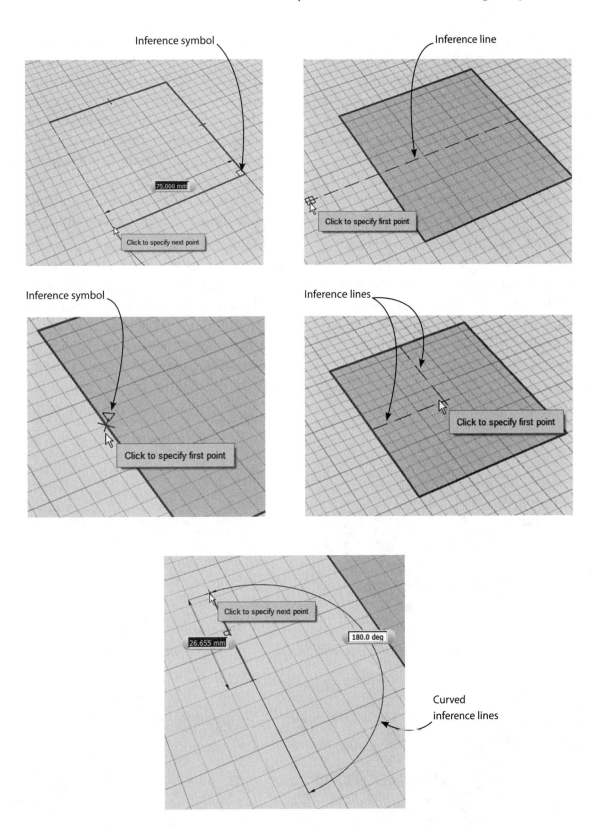

Figure 3-10 Inference lines and symbols alert you to specific lengths, positions, and angles.

Figure 3-11 shows how to edit the size of a closed profile sketch. Click its outline to bring up a dimension line and gear icon. Click the gear, select Edit Dimension from the context menu that appears, and click the outline again. Move (don't drag) the mouse, and the dimension line will move, too, enabling you to place it elsewhere. Click it in place and then click the dimension number. A text field will open; over-type the number in it and press ENTER. The sketch will adjust to the new size.

Some modelers like to start their sketches at the origin (point 0,0,0) because it makes counting grid squares easier. Another reason for starting there is that if you import your sketch into another program as part of a larger workflow, placing it at the origin will make it easier to find in that program.

When sketch lines overlap, they create a new, separate shape. Figure 3-12a shows a form extruded from the overlap of two circles. If you want to remove the overlapping lines, hover the Trim tool over the line you don't want until it turns red and then press DELETE (see Figure 3-12b). You can trim any sketch by drawing another line over it and trimming the overlap. Remember to first click the new line somewhere on the existing sketch so that both will be on the same plane. An inability to trim lines is usually because the sketches are on different planes.

Multiple sketches on the same plane cannot be moved or deleted independently. Moving one moves all; deleting one deletes all. However, the Trim tool can trim a whole sketch, which effectively deletes it. Click the Trim tool on the sketch, move the mouse away, and then move the mouse back over it and hover/move the mouse until the sketch turns red (see Figure 3-12c). Then you can click to delete.

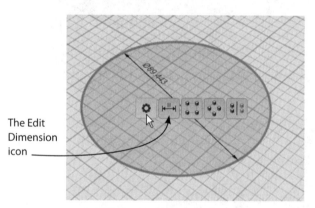

The Edit Dimension icon

1. Click the gear icon and select the first icon.

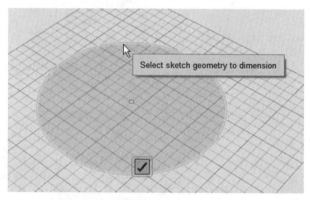

2. Click the outline.

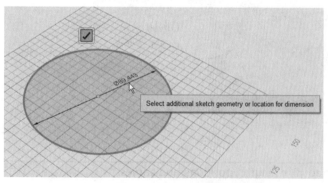

3. Move the dimension line that appears.

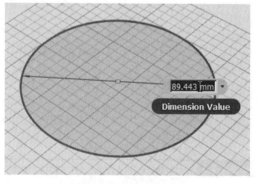

4. Overtype the numbers.

Figure 3-11   Edit a sketch's size by clicking it and selecting Edit Dimension from the context menu.

(a)

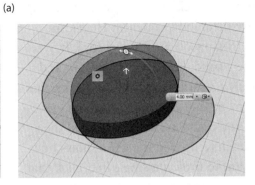

(b)

1. Click trim.

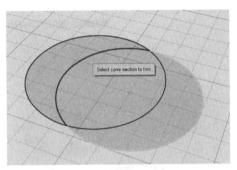

2. Hover the mouse over this line until it turns red and then press Delete.

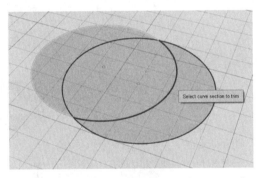

3. Hover the mouse over this line until it turns red and then press Delete.

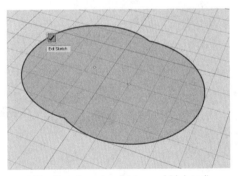

4. The lines have been trimmed (deleted).

(c)

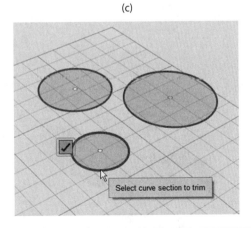

Figure 3-12 (a) Overlapping sketch lines create a separate shape. (b) Trim overlapping lines with the Trim tool. (c) Use the Trim tool to delete a sketch that's on the same plane as other sketches.

Solids are not associative to sketches, meaning you can move them independently. Here's an opportunity to show off the Project tool. If you want to move a solid, keep its sketch on it, and simultaneously preserve the locations of sketches around it (remember, same-plane sketches move as a group). You could move the solid and then project its sketch to it—Figure 3-13a shows how. If you just want to project a single line, click that line and then click the surface to project it to.

The Project tool is also useful for making a sketch from a complex solid that has evolved from multiple operations. Such a task might be impossible with the simple sketch tools available. Click Project, click a highlighted face of the solid, and then click again. The face will project a sketch of itself. That sketch can then be moved aside and altered as needed (see Figure 3-13b).

(a)

1. Click anywhere on the cluster of sketches.

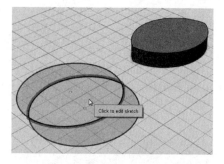

2. Click on the specific sketch to be projected.

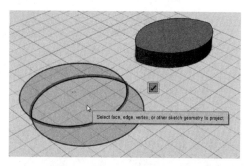

3. Click on the solid.

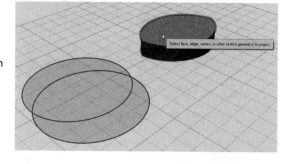

4. The sketch projects to the solid.

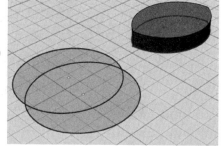

(b)

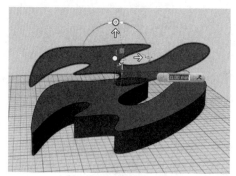

Figure 3-13 (a) Steps for projecting a sketch onto a solid. (b) Make a sketch of an existing solid face by projecting the face onto itself. Then move that sketch off the solid.

The Extend tool lengthens a line to intersect another line. To make Line A reach Line B, click B first and then click A. Line B's length will extend. This results in one continuous line instead of two line segments (see Figure 3-14).

Sketches are either closed profile or open profile. A *closed profile* has endpoints that are all connected and on the same plane (think multiple points on a sheet of paper). The darkened interior, or *face,* is evidence of a closed profile.

An *open profile* sketch does not have a face. This is caused by nonconnecting endpoints and/or endpoints on different planes. Many operations cannot be performed on open profile sketches.

The Text tool, accessed through the T on the menu bar, makes closed-profile letters. It uses your system's fonts, and independent operations can be performed on each letter (see Figure 3-15). Click it onto the workplane, click again on the location you want it, type in the dialog box

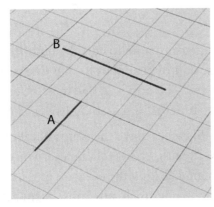

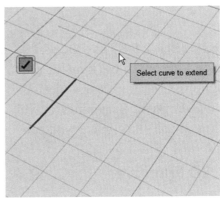

1. Click line B.

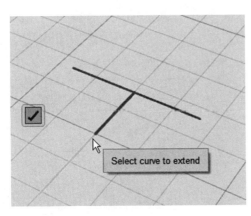

2. Click line A.

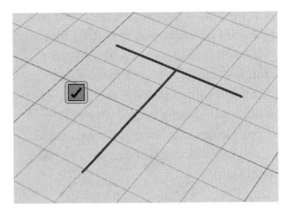

3. Line B's length will extend.

Figure 3-14 The Extend tool lengthens a line to intersect another line.

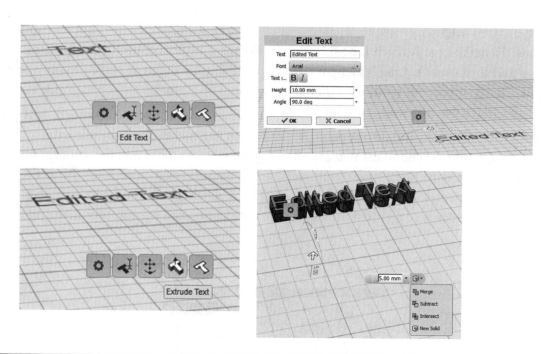

Figure 3-15 The Text tool lets you type text onto the workplane. Click the text to edit it. Here "Text" was changed to "Edited Text."

that appears, and then click OK. To edit, click on it again. A gear will appear. Click on the gear to see editing options. Copy the context menu by typing CTRL-C and CTRL-V. A manipulator will appear; move the copy off the original with it.

Sketches vs. 2D Primitives

The Primitives menu has 2D shapes that are also called "sketches" and have the same shapes as the ones in the Sketch menu. However, there are some significant differences. A Primitives menu sketch can be cruised to a solid and snapped to the center of a face. A Sketch menu shape cannot. Primitives don't overlap, and a shape between their overlapping parts does not form a new shape. However, a Primitives menu sketch and a Sketch menu sketch will overlap and result in a distinct shape.

The process for size-editing a Primitives menu sketch is similar to that of a Sketch menu sketch: click an edge, select Edit Dimension from the context menu, move a dimension line out, click it in place, and then click the

dimension note to edit the number. However, the edit only applies to that line (see Figure 3-16). The opposite line must be edited separately if a matching length is needed.

Sketches make the solids harder to see and work with, so hide them after you're done with them. Clicking the eye icon in the navigation panel hides them all. Subsequent sketches will be visible and need to be hidden. To just hide

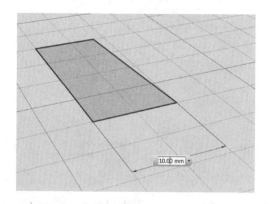

Figure 3-16 This Primitives menu sketch started as a rectangle. Its shape changed when one line was edited, because edits only affect the line they're applied to.

one sketch, select it (make sure you select both its face and outline), right-click, and then click the eye icon from the context menu. If you're sure you won't need a sketch anymore, select and delete it. You can hide solids and imported .stl files individually, too.

Sun Sculpture Wall Decoration Project

1. Sketch a circle by clicking the Circle sketch tool. See Figure 3-18 for the steps. You're prompted to choose a sketching surface; click anywhere on the sketch plane. Next you're prompted to choose the circle's center location—and choose its diameter. Type **3** for the diameter and press ENTER. A circle shape appears. Click the check mark to exit the sketch.

Figure 3-17 Sun sculpture wall decoration

2. Sketch a sunray with the Spline tool, which creates curved lines between points (see Figures 3-19 and 3-20). Click it anywhere on the circle to put the spline catch on the same plane. Then choose the spline's first point,

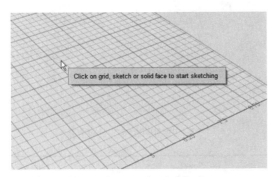
1. Click anywhere on the sketch plane to choose a sketching surface.

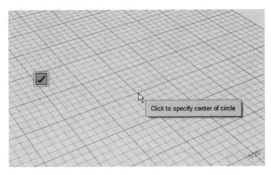
2. Click to choose the circle's center location.

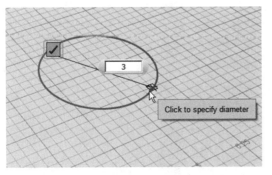
3. Type 3 for the circle's diameter and press Enter.

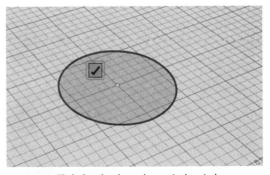
4. Click the check mark to exit the circle.

Figure 3-18 Making a circle with the Circle sketch tool.

subsequent points, and the ending point. Note the control points that appear (small, open circles). You can drag those with the mouse to tweak the shape (see Figure 3-20).

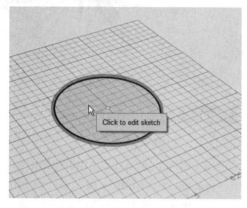

1. Click the spline tool anywhere on the circle to put the spline sketch on the same plane as the circle.

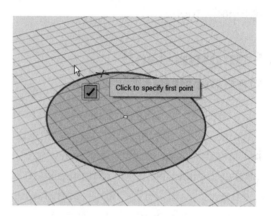

2. Choose the spline's first point.

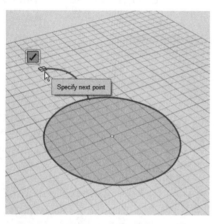

3. Choose the spline's subsequent points.

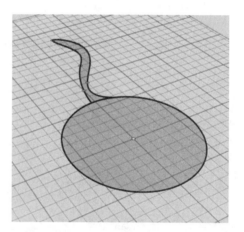

4. Choose the spline's end point.

Figure 3-19 Sketching a sunray with the Spline tool.

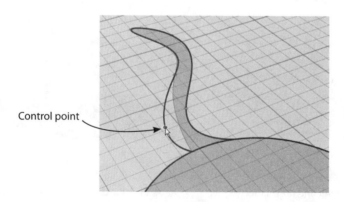

Figure 3-20 Drag the control points to tweak the shape.

3. Now select the sunray. Clicking once on the outline selects it; clicking twice on the face selects it and the outline. A gear icon appears; click it to reveal a context menu, as shown in Figure 3-21. Selecting just the sketch brings up fewer choices than selecting the face and sketch together. In this case, either selection works.

Although you can select tools from either the top-screen menu or open context menu, it's best to choose from the context menu because it shows options available for the immediate problem. Also, the top menu icons won't work if, for example, you're working on a sketch and the menu icon is only for solids.

4. Array the sunray. Select Circular Pattern from the context menu (see Figure 3-22). A glyph with two buttons appears: Sketch Entities and Center Point. Because the sketch is already highlighted, click Center Point (make sure the Full option from the drop-down arrow is selected). Click the mouse on the perimeter of the circle. Yellow arrows appear; drag them forward until you have the number of arrays you want and then click to set them.

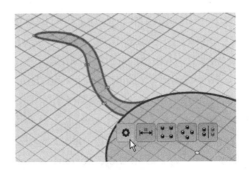

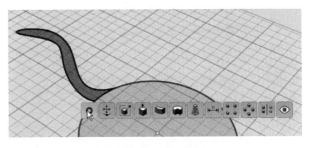

Figure 3-21 When you select the sunray a gear appears. Click it to see the context menu. More choices are available when the outline and face are selected than when just the outline is selected.

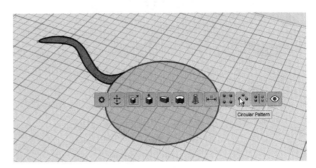

1. Choose Circular Pattern.

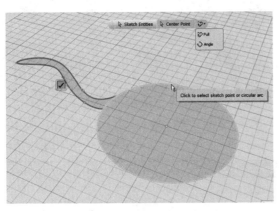

2. Click on Center Point. Then click the dropdown arrow and choose Full.

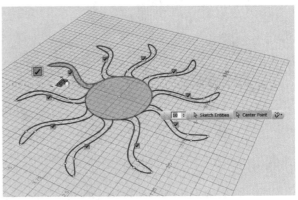

3. Drag the buttons until you have the number of copies wanted or type a number in the text field.

Figure 3-22 Array the sunray.

5. Extrude the sun by choosing Extrude from the Construction menu, selecting the entire sun with a selection window, and making sure New Solid is chosen from the drop-down menu (see Figure 3-23). Click the top face to select it. Once it's selected, it will darken, and an arrow will appear. Drag the arrow up. Type **.5** in the text field and press ENTER. You don't need the sketch of the sun anymore, so hide it.

6. Place a second, smaller circle on top of the sun (see Figure 3-24). Cruise a Primitives menu circle sketch into the canvas and snap it to the center of the sun. Type **1** for its diameter before clicking it in place. Then move it up half an inch.

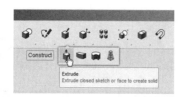

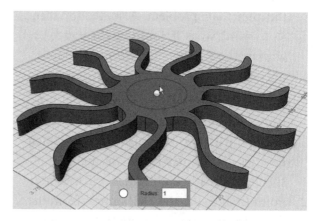

1. Cruise a Primitives sketch onto the sun and snap it to the center.

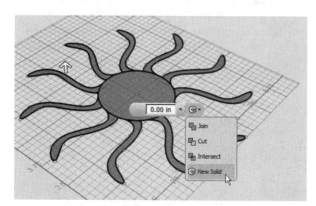

1. Select the entire sun, click the dropdown menu and choose New Solid.

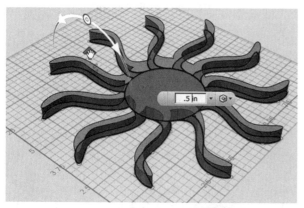

2. Drag the arrow up to extrude.

Figure 3-23 Use the Extrude tool to turn the sketch into a solid.

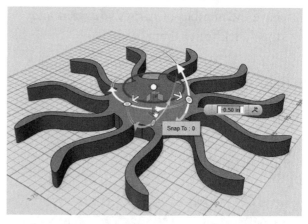

2. Make the sketch smaller and move it up.

Figure 3-24 Cruise a Primitives sketch onto the sun, size and move it.

7. Lofting fills the space between closed profiles. To loft the sun and circle, as shown in Figure 3-25, click Loft from the Construction menu and choose New Solid from the drop-down arrow. Then click separately on the sun and circle.

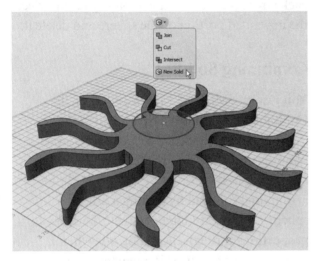

1. Choose the Loft tool's New Solid option.

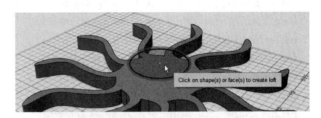

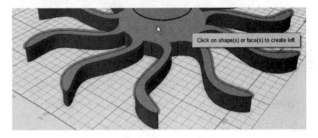

2. Click on both the sun and the circle sketch.

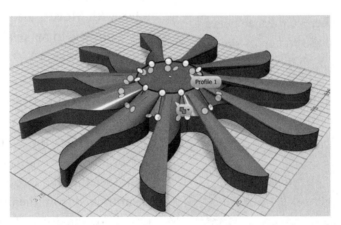

3. Click on the canvas to finish the loft operation.

Figure 3-25 Loft the sun and circle.

The Loft Tool

The Loft tool only works on closed profiles (they can be Primitives menu sketches or Sketch menu sketches). The faces must be selected one at a time (hold the SHIFT key down to make multiple selections), because lofting doesn't work with a selection window. Make sure you select both the outline and the face; if the tool doesn't work, you probably just selected the face. You will get different results based on how many faces are selected for one loft operation; for example, selecting them all will result in a different shape than selecting and lofting two profiles at a time. Experiment with different profile shapes, the number of profiles selected for one loft, and the position of the profiles. Profiles don't have to be aligned, but if a loft is impossible for the positions you chose, an error message will appear. Investigate Loft's other options: Join, Cut, and Intersect. You can make interesting, complex forms with the Loft tool.

8. Finally, let's add a sphere. Cruise a sphere to the sun and snap it to the center, as shown in Figure 3-26. Then push it halfway down. You're done!

Figure 3-26   Cruise and snap a sphere to the top.

The sun sculpture is thick enough that the sphere doesn't protrude through the bottom. But if it did, here's how to cut the protruding part off: From the Combine menu, select Merge (see Figure 3-27). A glyph with two buttons appears—Target Solid and Source Solid. The Target Solid button should be highlighted; if it's not, click it and select the sun. Next, click the Source Solid button and select the sphere. Press ENTER or just click anywhere on the canvas. Now you can select the unwanted portion of the sphere and delete it.

Combining Solids

You just saw what the Merge option of the Combine menu can do. Experiment with its Subtract and Intersect options, too. Subtract removes one shape from another and is handy for cutting holes in solids. Figure 3-28 shows how to make a bead by cutting a hole in a sphere. (Note that I changed the cylinder's proportions through the dialog box when I cruised it to the canvas. Remember to press the TAB key to change text fields. Then rotate it 90° with Move. I also set the View Cube to Orthographic to make aligning the the sphere and cylinder easier.)

Click the Combine menu and choose Subtract. Select the Target Solid button and click the sphere. Next, click the Source Solid button and click the cylinder. Then click the cursor anywhere on the canvas or press ENTER. You'll see that the second solid (cylinder) is subtracted from the first solid (sphere). Obviously the order of selection is important. A different order will give a different result.

Importing a Picture File

It's often quicker to import a picture to model from than to model from scratch. 123D Design can import .svg format files. These are files

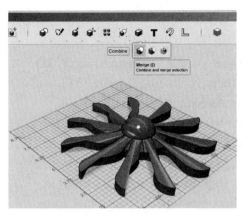

1. Select Join. The sphere is the Target.

2. The sun is the Source.

3. Click on the canvas to complete the operation, and then delete any protruding parts underneath.

Figure 3-27 Combine the sphere with the body and then delete the bottom half of the sphere to give the sculpture a flat back.

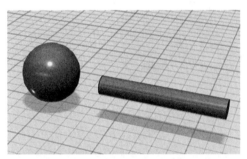

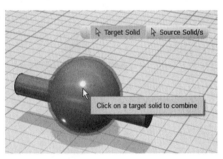

1. Cruise a sphere and cylinder onto the canvas and change the cylinder's proportions.

2. Click on the Target sphere.

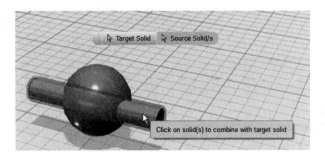

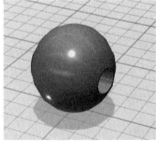

3. Click on the Source cylinder.

4. Click on Combine/Subtract and then click on the workspace to complete the combine operation.

Figure 3-28 Make a bead by subtracting a cylinder from a sphere with the Join menu's Cut option.

made by drawing programs such as Inkscape, a cross-platform, open source (free) one, or Adobe Illustrator. However, an importable file must be completely vector—no raster images in it at all—and have a transparent background. A "clean," carefully drawn picture works best. If the drawing imports as an open profile, it either has gaps in the line work or lines that aren't coplanar. A drawing that has a lot of short lines instead of fewer, longer ones will require more work to model. It may be easier to fix such problems in the original drawing program than in Design.

Star Cookie Cutter Project

The star-shaped cookie cutter shown in Figure 3-29 was made by importing the .svg file next to it. To start this project, set Units to millimeters

(mm). We'll use the Offset, Measure, and Scale tools. Here are the steps to follow:

1. Import the .svg file by going to the 123D Design menu, choosing Import SVG, and selecting As Sketch (see Figure 3-30). Navigate to the file on your computer and import it. SVG files usually have an Internet browser icon; here they appear as Internet Explorer icons. Once imported, the file becomes line sketches that can be manipulated like any other sketch. We don't know what size it is, so we'll just model it by proportion and scale to size when finished.

2. To offset the figure, click the Offset tool (see Figure 3-31). Click the sketch to put the offset lines on the same plane. Then click anywhere on the outline (the whole outline should highlight) and click on a location to place the offset lines.

Figure 3-29   An .svg file of a star and the cookie cutter made from it.

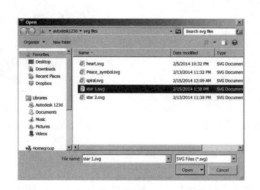
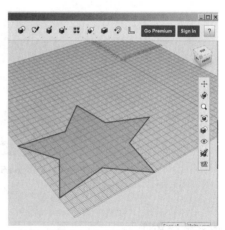

Figure 3-30   Import the .svg file into the canvas.

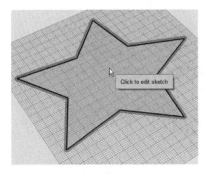
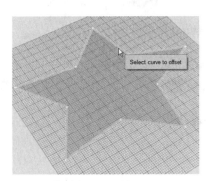
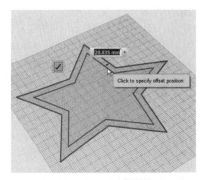

1. Click on the sketch.   2. Click on the outline.   3. Click on the location you wish to place the offset sketch.

Figure 3-31 Offset the star.

3. Double-click to select the new border. Make sure the Extrude tool's New Solid option is chosen so that we can extrude the outline (see Figure 3-32).

    Now we need to make this model an appropriate size for a cookie cutter. First, we'll measure it to see how big it is, and then we'll scale it down to about 3mm wide.

4. Measure the star by clicking the Measure tool and choosing the Face/Edge/Vertex option (see Figure 3-33). Click the mouse on opposite sides of the star and a pop-up box

appears. In the Distance field, we see the star is about 501mm wide. Too big!

5. Scale the star by selecting it and choosing Scale from the glyph that appears at the bottom (see Figure 3-34a). To make it 76mm, type **.15** in the factor box (76mm/501mm) and click the canvas.

6. Fillet the edges of the star by selecting a top edge and choosing Fillet from the context menu (see Figure 3-34b). Then hold the SHIFT key down and select all top perimeter edges so they'll fillet at the same time. Either drag

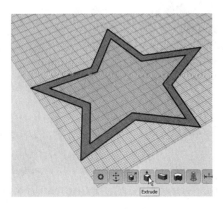
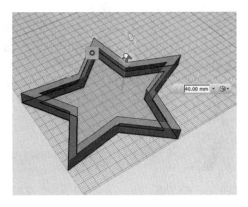

1. Select the border.   2. Extrude the border.

Figure 3-32 Select and extrude the new border.

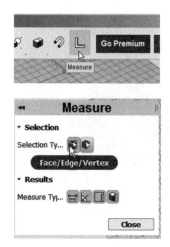 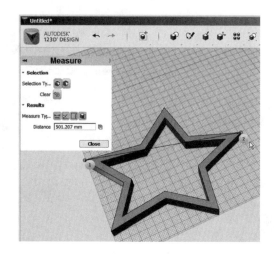

Figure 3-33 Activate the Measure tool and click two points to get the distance between them.

(a)

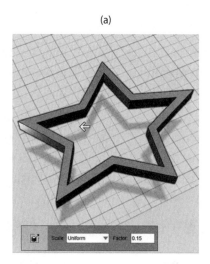

(b)

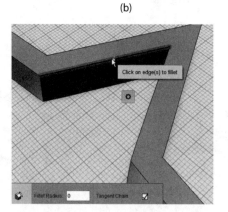 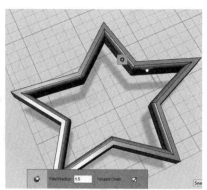

1. When the glyph appears, click Scale.   2. Select all the top interior perimeter edges.   3. Drag the fillet or type a number in the text field.

Figure 3-34 Scale and fillet the star.

the arrow and "eyeball" a fillet or type a number in the text field. Here, I typed **1.5**. Splash some color or texture on, and you're done!

Jell-O Mold Tray Project

We're going to use the star .svg file again, but this time we'll import it as a solid and use it to cut the shape in a tray of Jell-O molds (see Figure 3-35). In the process, you'll learn how to mirror, make a rectangular pattern, and extrude through a solid. To begin, set Units to inches and then follow these steps:

1. Draw a rectangle by clicking the Rectangle sketch tool onto the canvas (see Figure 3-36). Type **4** in one text field, press the TAB key, and type **6** in the next to make a 4"×6" rectangle. Then extrude it up 1.5".

2. Draw a mirror line (see Figure 3-37). Click the Polyline tool on the vertical face to orient a sketch plane. Click the line's first endpoint in the middle of the face. Click a second endpoint right below it.

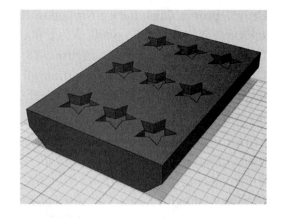

Figure 3-35 Jell-O mold tray. The star cutout is from the same file used for the cookie cutter.

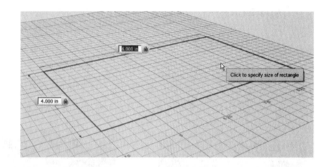

Figure 3-36 Start the Jell-O mold with a 4"×6" rectangle.

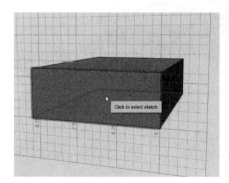

1. Click a polyline onto the vertical face.

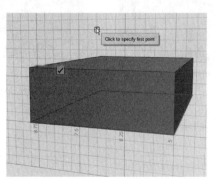

2. Click the polyline's first point.

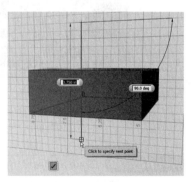

3. Click the polyline's second point.

Figure 3-37 Draw a mirror line.

3. Draw a profile. Click the mirror line so that subsequent lines will go on its sketch plane. Draw a triangle as shown in Figure 3-38. Don't just draw one angled line; draw all three. Then extrude the triangle through the tray.

4. Mirror the profile (see Figure 3-39). Select it and choose the Mirror icon from the context menu. With the glyph's Sketch Entities highlighted, select all three of the triangle's lines. Then highlight Mirror Line on the glyph, click the mirror line drawn in step 2, and click anywhere on the canvas. The profile will appear on the opposite side. Extrude it.

> The mirror command for a sketch needs to be called up from the context menu, because the mirror command in the menu bar only mirrors solids. A mirror line must be in the same plane as the sketch. If you copy a sketch (CTRL-C, CTRL-V), you cannot mirror both sketches around one line. You must make a mirror line for each sketch. Make sure that you select a shape's outline, not just its face, when mirroring.

5. (Optional) If the mold looks too thick, highlight its top face and choose Tweak from the pill. Then press/pull it down (see Figure 3-40).

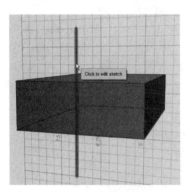

1. Click on the mirror line.

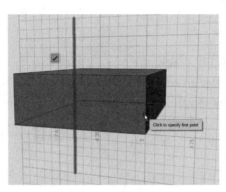

2. Click the triangle's first point.

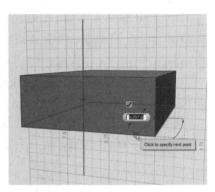

3. Click the triangle's second point.

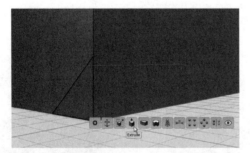

4. Select the triangle and choose Extrude from the context menu.

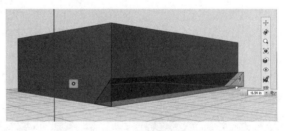

5. Extrude the triangle through the tray.

Figure 3-38 Draw and extrude a triangular profile.

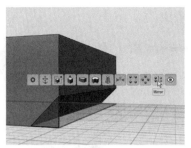

1. Select the triangle and choose Mirror from the context menu.

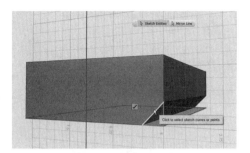

2. With Sketch Entities highlighted, select the triangle's lines.

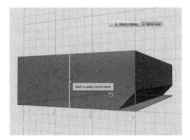

3. With Mirror Line highlighted, select the mirror line.

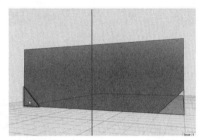

4. Click anywhere on the canvas to place the mirrored triangle.

5. Extrude the triangle through the tray.

Figure 3-39 Mirror and extrude the profile.

Figure 3-40 Tweak the mold by press/pulling it down, if desired.

6. Import, measure, and scale the star .svg file (see Figure 3-41). This time we'll import the .svg as a solid. It enters with its own sketch and very big. Click the Measure tool. This brings up a dialog box. There are two Selection Type icons: the first is Face/Edge/Vertex and the second is Body. Click the first icon and then click the cursor on two endpoints. The distance between them appears in the text field. Note the four icons near the bottom: Distance, Angle, Area, and Volume. You can click each for these specific measurements. To remeasure, click Clear; when you're finished, click Close.

1. Import the .svg file as a solid.

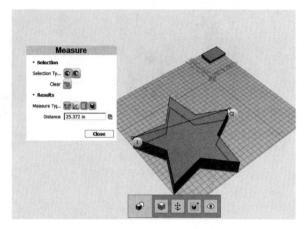

2. Measure the solid.

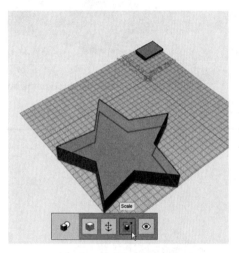

3. Select the solid and choose Scale.

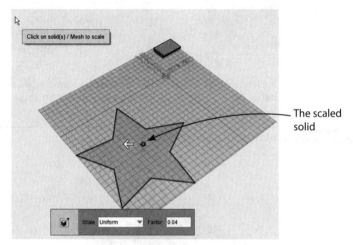

4. Scale the solid either by dragging it or by typing a number in the text field. Since the outline wasn't selected, it didn't scale with the solid.

Figure 3-41 Import, measure, and scale the solid star.

This star is 25 inches wide, so select it to bring up the glyph at the bottom. Choose Scale and then either drag it smaller with the arrow or type in a scale factor. To make the star one inch wide, type **.04** in the Factor box (1/25). In Figure 3-41, only the solid was highlighted, so only it scaled; if the sketch was highlighted, it would have scaled, too. But you can hide or delete the sketch because we don't need it.

7. Snap the star to the mold (see Figure 3-42). With the Snap tool, click the star and then on the tray. Move the star precisely where wanted and then push it down a bit to intersect the tray. The exact amount you push it down will be the depth of the mold.

8. Array the star (see Figure 3-43). Choose Rectangular Pattern from the Pattern menu.With the Solid/s button in the glyph highlighted, select the star. Then highlight the Direction/s button and click a tray edge. Two arrows will appear by the star; drag one arrow left and the other arrow down to make equally spaced copies. You can also type the number of copies wanted in the text field.

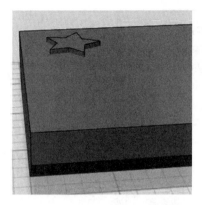

Figure 3-42 Snap the star to the mold and then move it down to intersect.

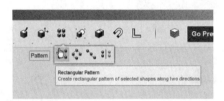

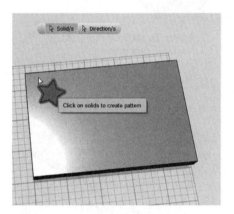

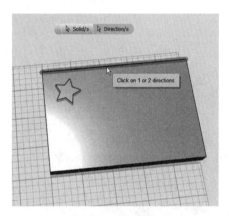

1. Select the star.

2. Click a tray edge.

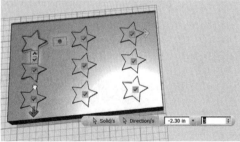

3. Drag the button left until you have the number of stars wanted.

4. Drag the button down until you have the number of stars wanted.

Figure 3-43 Copy the star with the Rectangular Pattern.

When you use CTRL-V and CTRL-P to copy a sketch, the copy is put on a different sketch plane. If you want them on the same plane, a way to achieve that is to array them. All arrayed copies are on the same sketch plane.

9. Cut the star shape out (see Figure 3-44). Choose Subtract from the Combine menu. With the glyph's Target Solid button highlighted, click the tray. Then highlight the Source Solid/s button and click a star. Click anywhere on the canvas to finish; the tray will subtract the star, resulting in a cutout. Repeat, but this time select the rest of the stars at once (hold the SHIFT key down). Done!

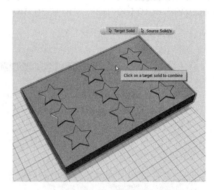

1. With Target Solid highlighted, click the tray.

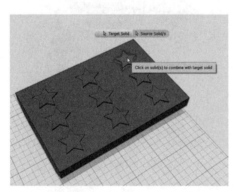

2. With Source Solid highlighted, click a star.

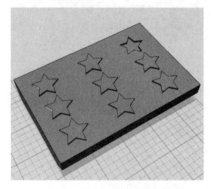

3. Click anywhere on the canvas to finish the operation.

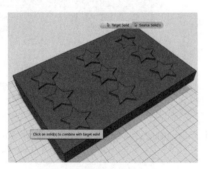

4. Repeat steps 1 and 2, but select all the stars.

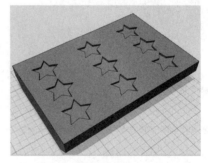

5. Click anywhere on the canvas and all the stars will be cut out.

Figure 3-44   Cut the star shapes out.

10. (Optional) If you want to make the molds deeper, highlight them and choose Press/Pull from the context menu. Then push them down a bit (see Figure 3-45). Okay, now we're done!

Peace Sign Pendant Project

In this project we'll import an .svg file of a peace symbol and turn it into a necklace pendant (see Figure 3-46). In the process we'll offset, trim, group, and use the Orthographic view. To begin, set Units to mm and then follow the steps shown in Figure 3-47.

Figure 3-45 Select and press/pull the stars down to deepen the molds.

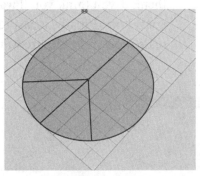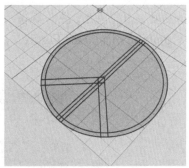

Figure 3-46 An .svg file of a peace sign and the model made from it.

1. Import the .svg sketch.

2. Offset the sketch's lines.

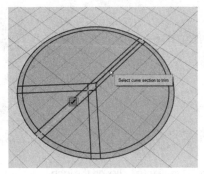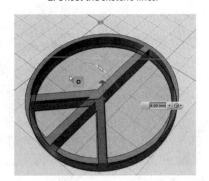

3. Remove unwanted lines by trimming them.

4. Extrude the sketch.

Figure 3-47 Import the .svg file and offset its lines to give the symbol thickness.

1. Import the file

2. Offset the file's lines. First click on the imported sketch so that the offset lines get placed on the same plane. Second, click the line to offset. Third, click the place to offset.

   If you press ENTER after offsetting a line, you'll have to return to the Sketch menu to choose Offset again. But if you click the canvas, the tool remains active, which is a faster way to offset numerous lines.

3. Trim the lines you don't want. Click the Trim tool (in the Sketch menu) onto the sketch. Then click the specific lines to remove.

4. Extrude the sketch.

5. Add a loop to the pendant (Figure 3-48). Rotate the symbol upright with the Move tool. Then bring a torus into the canvas. Before snapping it on the symbol, type the desired size in the text fields (remember to press TAB to toggle between them). Then move the torus up and back. Set the ViewCube to Orthographic to make the alignment easier to see.

6. Combine the pendant and loop by choosing Merge from the Combine menu (see Figure

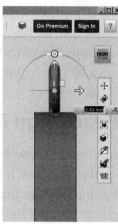

Figure 3-48   Add a loop to the pendant by snapping a torus to its face and adjusting the torus size.

3-49). With the glyph's Target Solid button highlighted, click the peace sign. Then highlight the glyph's Source Solid/s button and click the loop. This welds them together permanently, as evidenced when you run the cursor over one or the other: both pieces highlight. Alternatively, select and group them, if you think you'll want to take them apart later. Finally, measure the pendant and scale it down to the size wanted.

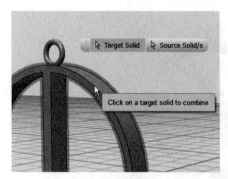

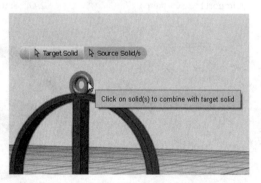

1. With Target Solid highlighted, click on the pendant.

2. With Source Solid/s highlighted, click on the loop.

3. Click anywhere on the canvas to complete the operation.

Figure 3-49   Weld the pendant and the loop together with Combine | Merge.

Personalized Business Card Project

To model the personalized business card shown in Figure 3-50, we'll sketch with the Fillet tool. To begin, set Units to mm and then follow these steps:

1. Use the Polyline tool to sketch a rectangle. Then curve the corners with the Fillet tool (see Figure 3-51). Click Fillet first on the rectangle to put the fillet lines on the rectangle's sketch plane. Then click the fillet's two endpoints. Note the arrow that appears. You can drag it to tweak the fillet.

Figure 3-50   A personalized business card.

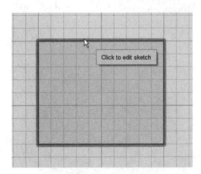

1. Sketch a rectangle with the Polyline tool.

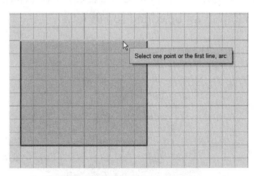

2. Click the Fillet tool on the sketch and then on the first endpoint.

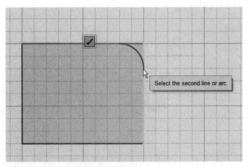

3. Click the Fillet tool on the second endpoint.

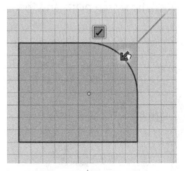

4. If needed, drag the arrow to tweak the curve.

Figure 3-51   Sketch a rectangle with curved corners.

2. Offset and extrude the rectangle (see Figure 3-52). Click the Offset tool on the rectangle's perimeter, offset the distance desired, and click to place. Extrude the perimeter and interior different heights.

3. Click the Text tool once on the card to select the sketch to place it on, and twice to select the location (see Figure 3-53).

   The Text dialog box will appear. Type what you want and then click OK (Figure 3-54).

4. Activate the Text tool by clicking on the T. Then click on the the sketch plane and click again on the text location. Type in the dialog box that appears, then click OK to exit .

5. Click on the text to bring up the gear; then click the gear to open the context menu. Click the Move tool and position the text. Then click the gear again to access the Extrude tool. Pull the text up, and you're done (Figure 3-55). Once the text is extruded, each letter can be edited separately.

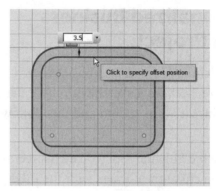

1. Offset the perimeter.

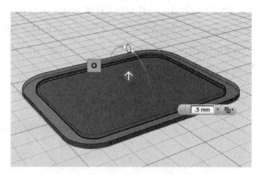

2. Extrude the perimeter.

Figure 3-52 Offset and extrude the rectangle.

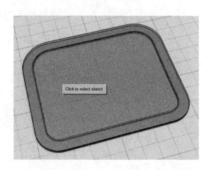

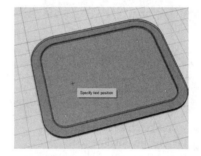

Figure 3-53 Click the sketch plane and then click where to start the text.

Figure 3-54 Type in the dialog box.

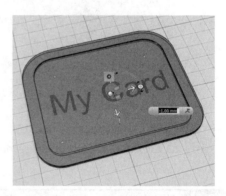

Figure 3-55 Position the text before extruding it up.

Pac-Man Project

In this project we'll use the Arc, Revolve, Shell, and Split Solid tools to model the Pac-Man and Pinky characters in Figure 3-56. While doing so, you'll learn how to save them to, and access them from, your online 123D account. We'll start with Pac-Man. To begin, set Units to mm and follow these steps:

1. Cruise a sphere onto the canvas (see Figure 3-57). Keep its default 10mm radius and click in place. Set the ViewCube to Top and Orthographic; these settings will make it easier to sketch.

2. To draw the mouth, we'll need a profile of an arc (see Figure 3-58). Draw a polyline and then activate the Three Point Arc tool. Click it first on the grid, second on one endpoint, and third on the other endpoint. Drag the arc out and click to set.

Figure 3-56 Pac-Man and Pinky.

Figure 3-57 Cruise a sphere onto the canvas and change the ViewCube settings to Top and Orthographic.

1. Draw a polyline. Then click the Three Point Arc tool on that polyline.

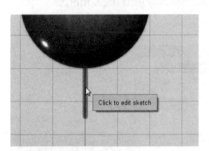

Click to edit sketch

2. Click the Three Point Arc tool on one endpoint.

Click to specify start point of arc

3. Click the Three Point Arc tool on the second endpoint.

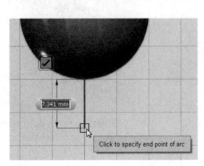

7.341 mm

Click to specify end point of arc

4. Drag an arc.

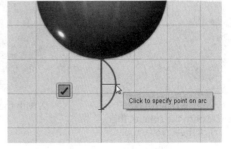

Click to specify point on arc

Figure 3-58 Sketch a profile using the Polyline and Three Point Arc tools.

3. Mirror and move the profile as shown in Figure 3-59. Mirror the arc around the polyline and then delete the polyline. Use the Move tool to rotate the profile 90°, raise it, and move it about halfway into the sphere.

4. Rotate the profile (see Figure 3-60). To do this, first cruise a 2D Primitive circle onto the canvas and snap it to the sphere. You need a primitive because an ordinary sketch won't snap. From the Construct menu, choose Revolve. A two-button glyph will appear. Click the Profile button and select the circle. Then click Axis. Another glyph will appear; type **90** in its text field to rotate the profile 90° and then select Cut from the drop-down menu. Select the circle's perimeter as the axis of revolution. A red volume should appear, indicating the cut just made. Click the canvas to finish the operation and then delete the profile.

Different profiles make different shape grooves. The groove's shape is also affected by how far you intersect the profile into the solid. Before deleting the profile, you might trim its top line, sketch a different shape line on it, and see what shape groove it makes. Edit the size of the profile and try again. Experiment with different profiles. Remember, you can undo as far back as needed.

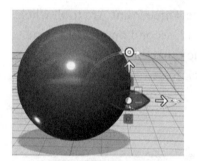

1. Mirror the arc around the polyline.

2. Move the profile into place.

Figure 3-59 Mirror the profile and then move it up and about halfway into the sphere.

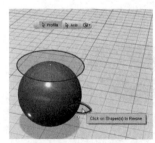
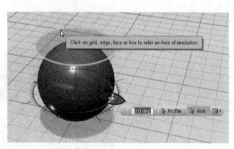
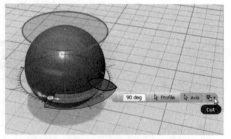

1. Cruise and snap a Primitive circle onto the sphere. Then click the Revolve tool.

2. With the Profile button selected, click on the circle.

3. With the Axis button selected, type 90 to rotate the profile 90°.

Figure 3-60 Revolve the profile around the sphere.

5. Finesse the groove (see Figure 3-61a). Select the planes and press/pull up or down. Select the edges and fillet. Add two spheres for the eyes (see Figure 3-61b). Change their radius to 2mm before clicking them onto the canvas, and then move them into the large sphere.

Add colors and then group everything. Grouping is needed because Combine won't preserve separate colors. When two solids are combined, the source solid becomes the color of the target solid.

Uploading a File to Your 123D Account

Let's save Pac-Man to your online 123D account now and get acquainted with its features. Click Save from the 123D App menu and choose To My Projects (see Figure 3-62). The model is then spirited off to your 123D account (if you're not already logged in to your Autodesk account, you'll be prompted to do so). When the model gets there, you'll be asked to name and describe it, and have the option of publishing it publicly

1. Finesse the groove by selecting its planes and edges and using Press/Pull and Fillet on them.

2. Cruise two more spheres out, size them, snap them to the larger sphere, and push them into it.

Figure 3-61   Turn the sphere into Pac-Man with a mouth and eyes.

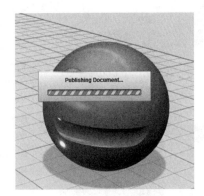

Figure 3-62   Saving a file online with the To My Projects option.

or privately. Public is the default and puts your model in the 123D Gallery. There, anyone can see, like, and download it. Click the drop-down arrow to find the Private option, which allows only you to see it. You can change it to public later, as well as make a public file private. Then save. Click on Models in the left sidebar to see the file's thumbnail, as well as thumbnails of other uploaded models.

You can also access your account through the Sign In menu (where your name now appears). Choose My Projects (see Figure 3-63). This will take you to the same place.

Note that each thumbnail has a gear in the upper-left corner. Click it for an options menu. Here, you can edit and delete the model, send it to other apps, or send it to a third party fabricator (see Figure 3-64). The red button next to the gear icon indicates the model is private. Clicking on the thumbnail itself brings up a window with an Edit/Download button. Click that button for a menu that has an option for downloading files of the model.

Figure 3-63 Access your online account and files via the Sign In menu.

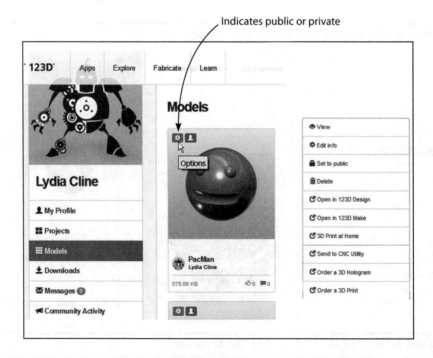

Figure 3-64 The Models page has a thumbnail of each uploaded model, and each thumbnail has an options menu.

Model vs. Project

You may be wondering what the difference between a model and a project is. A *model* is just one file, whereas a *project* is a collection of files, typically of different formats. For example, you can upload a 123D Design model and then upload relevant .dwg, .stl, .obj, and .pdf files to go with it. You can also upload non-Autodesk software files. The Project page has a Create a Project button to start this process. If the project is public, all of it is available for anyone to download. When you make a file public, you can choose from different license types (these govern public use).

Pinky Project

Open a new 123D Design file, and we'll model Pinky. Set the Units to mm and follow these steps:

1. Model the body (see Figure 3-65). Cruise a cylinder and sphere onto the canvas, keeping their default sizes. Move the sphere halfway down into the cylinder. Then cruise a cone in, setting its radius to **3** and its height to **6**. Intersect it halfway into the cylinder. This cone will be used to make the ruffle.

2. Array and Join the cone by using the Circle Pattern tool and slightly overlapping the copies, as shown in Figure 3-66. From the Join menu, choose Subtract. For Target Solid, click the cylinder; for Source Solid, drag a selection window around all the cones. Click the canvas to finish.

3. Hollow out the cylinder (Figure 3-67). Select each edge of the ruffle and fillet it 1mm. Next, select the bottom plane and choose Shell from the gear. The cylinder will immediately become hollow. Keep in mind that you can't Combine | Subtract a shelled solid.

1. Cruise a cylinder and sphere onto the canvas.

2. Move the sphere into the cylinder.

3. Cruise a cone into the canvas and intersect it with the cylinder.

Figure 3-65 Model Pinky's body with a cylinder and sphere. The cone will be used for the ruffle.

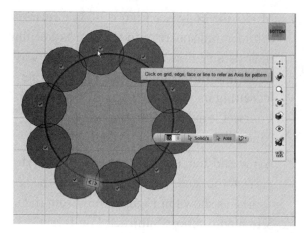

1. Array the cone.

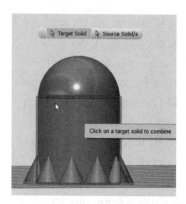

2. The cylinder is the target.

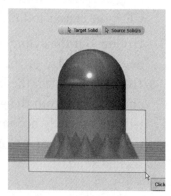

3. The cones are the source.

4. Click on the canvas to subtract the cones from the cylinder.

Figure 3-66 Array the cone. The result should look like the last graphic.

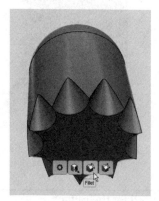

1. Fillet

2. Shell

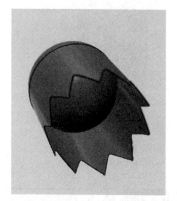

3. Result of the shell operation

Figure 3-67 Hollow out the cylinder with the Shell tool.

4. Let's remove the bottom of the sphere to see how the Split Solid function works. Cruise a box onto the canvas and align it with Pinky. Click the box's front to activate a vertical sketch plane. Then draw a polyline where the sphere and cylinder intersect (see Figure 3-68). We'll split the sphere along this line.

   From the Modify menu, choose Split Solid. A two-button glyph will appear. Highlight Body to Split and select the sphere. Then highlight Splitting Entity and select the line. A red plane will appear, showing where the solid will be split (see Figure 3-69). Click the canvas to finish. The sphere is now cut in half; you can delete the bottom.

5. To apply color, open the Materials menu and make the body and eyes different colors (see Figure 3-70). When applying color, remember to highlight the part of the model to color and to click both the outer ring and inner diamond. Then check the Apply Overlay box.

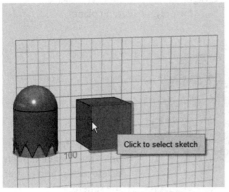

1. Cruise a box into the canvas and click on its vertical face as the sketch plane.

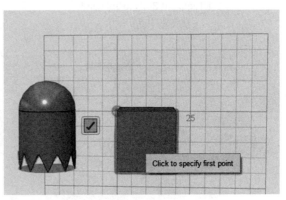

2. Click the polyline's first point.

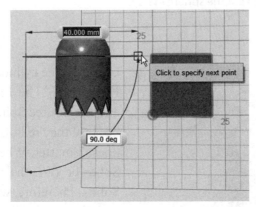

3. Click the polyline's second point.

Figure 3-68 Draw a line to split the sphere with.

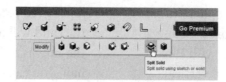

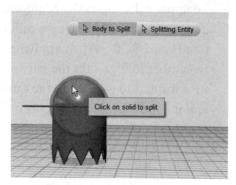

1. Highlight Body to Split and click on the sphere.

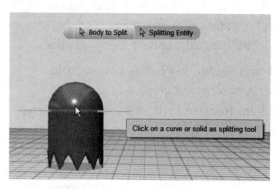

2. Highlight Spliting Entity and click on the polyline (cutting plane).

3. This will appear.

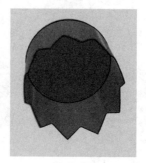

4. Click on the canvas to finish the operation. Then delete the bottom part of the sphere.

Figure 3-69 Use the Split Solid tool to cut the sphere in half.

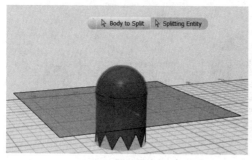

Figure 3-70 Apply separate colors to the body and eyes.

6. Combine the cylinder and sphere as shown in Figure 3-71. The Target Solid is the cylinder, and the Source Solid is the sphere. However, because they're both the same color, it doesn't matter. When combined, Source Solids become the same color as Target Solids. Therefore, we won't combine the eyes with the rest of the model. So just group the eyes with the now-combined body. Display the model as Materials Only so its outlines won't show.

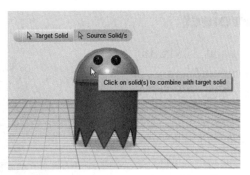
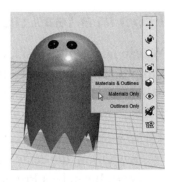

Figure 3-71 Combine the cylinder and sphere. Then group the combined model with the eyes. The last graphic shows the model displayed as Materials Only.

Inserting a File into an Open File

Time to bring Pac-Man back into the picture. At the 123D Apps menu, choose Insert (see Figure 3-72). If you're not signed into your Autodesk account, you'll be prompted to do so. The Projects page will load. Note the tabs at the top to browse your computer and the Gallery. Select the Pac-Man file and click the Open Selected Project button (or just press ENTER). Pac-Man will load into the Pinky file.

1. Sign into your 123D account. Pinky will wait.

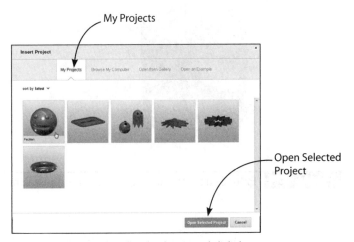

2. Find Pac-man's thumbnail, select it, and click the Open Selected Projects button.

3. Pac-man will load into the Pinky file.

Figure 3-72 Bring one Design file into another via the 123D Apps menu's Insert option.

Frog in a Tea Cup Project

We can import files besides .svg and .123dx. In this project, we'll model a tea cup and import an .stl file of a frog to put in it (see Figure 3-73). Set Units to mm and then follow these steps:

1. To model the cup's body, draw a circle sketch down using the default radius (see Figure 3-74). It can be a Primitives menu circle or a Sketch menu circle. Extrude it up 27mm and then taper it with the tweak manipulator on top. That manipulator only appears at the initial extrusion.

2. Hollow the cup out using the Shell tool and drag a 2mm thickness (see Figure 3-75).

Figure 3-73   An .stl file of a frog lurks at the bottom of this tea cup. The material applied is clear glass.

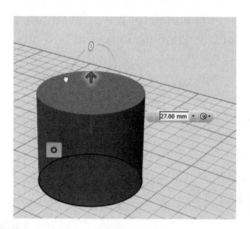

Figure 3-74   Extrude a sketch and taper it with the tweak manipulator.

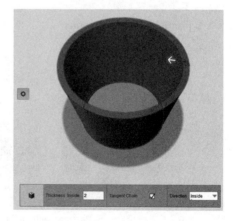

Figure 3-75   Hollow the cup out with the Shell tool.

3. Fillet the top 2mm and the bottom 5mm (see Figure 3-76). Hide the circle sketch at the bottom, because you won't be able to fillet over it, and hide the grid if it's in your way. Select both edges of the top to fillet.

4. Draw a handle (see Figure 3-77). Cruise a box next to the cup, and with the Spline tool, sketch a handle on the box's vertical plane. Next, move the handle over to intersect it with the cup. It's easiest if you set the ViewCube to Front and Orthographic. Then cruise a Primitives menu circle sketch to the cup and position it at the end of the spline.

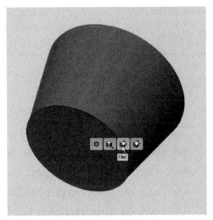

1. Select the bottom.

2. Fillet the bottom.

3. Select the top.

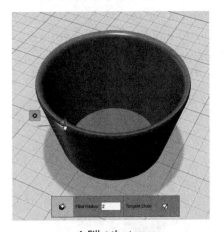

4. Fillet the top.

Figure 3-76 Fillet the cup's top and bottom.

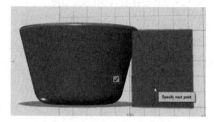

1. Click on the box's vertical sketch
   plane and draw a spline.

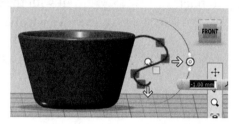

2. Move the spline next to the cup.

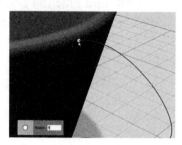

3. Snap a Primitive circle
   sketch to the cup.

Figure 3-77   Sketch a handle.

5. Use the Sweep tool to turn the handle sketch into a solid (see Figure 3-78). Activate and set the Sweep tool to New Solid. Click the circle sketch for Profile and the handle sketch for Path. Then click the canvas. Move and intersect the handle with the cup. You may need to activate the Move tool from the Transform menu at the top of the screen. Weld the cup and handle together with Combine | Merge.

6. Now we just need to import the frog. At the 123D menu, click Insert. After the online window loads, click Browse My Computer (see Figure 3-79). Navigate to the .stl file (note in the lower-right field all the file types that can be imported) and either click Open or press ENTER. The .stl file will enter at the origin.

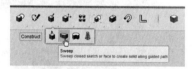

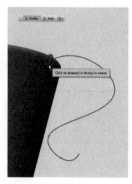

1. Select the circle.

2. Select the path.

3. Click on the canvas
   to finish.

Figure 3-78   Use the Sweep tool to turn the handle sketch into a solid. Then move the handle to intersect it with the cup.

 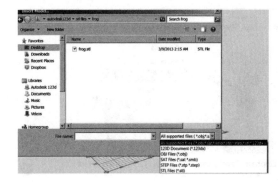

Figure 3-79 Import the frog .stl file.

7. The frog is much bigger than the cup, so select and scale it down (see Figure 3-80). Then scale both it and the cup up to whatever size you want. Apply a material of your choice, combine or group the cup and frog, and you're done!

A large file may take a while to load. It may also be tiny or gigantic relative to your open file and therefore require some zooming in or out to find. Both .stl and .obj files can be scaled, rotated, moved, colored, and hidden (they are the "mesh" files referenced by the navigator panel's eye icon). Some can be modified with the Combine menu's tools. If combining won't work on a particular file, a dialog box will advise

you to take it to the Meshmixer app. Know that while in Design, the changes you make to the stl model are really just made to the invisible shell around it. Meshmixer will take you into that shell to modify the mesh itself.

I bet you're wondering where you can get a cool frog like this one. Two places to check out are the123D Gallery at www.123dapp .com/Gallery/content/all and Thingiverse at www.thingiverse.com. Remember that the 123D Gallery has Premium models made by professionals (see Figure 3-81). Free members can download 10 each month; Premium members can download an unlimited amount. Alternatively, you can make your own in 123D Creature.

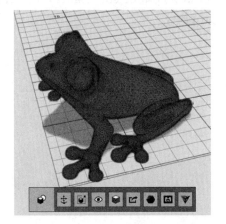 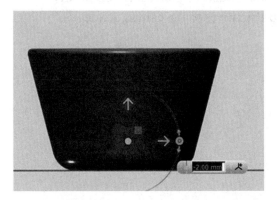

1. Scale the .stl file.   2. Move the .stl file into the cup.

Figure 3-80 Select and scale the frog as needed.

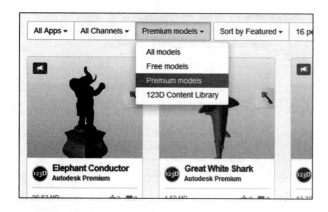

Figure 3-81   Premium models are available in the 123D Gallery.

Browse other people's models for ideas and inspiration. Download and study how they made them. If you want to use them for your own purposes, note that most have a Creative Commons (CC) license, which is a public copyright. CCs grant you, the downloader, varying degrees of rights. For instance, the modeler may or may not allow commercial use, require attribution, or forbid derivative works. The terms will appear when you click the Download button. Upload and strut your own stuff on those websites, too!

Summary

In this chapter we used 123D Design Desktop's tools to create models of increasing complexity. We cruised Primitives to the canvas, made

sketches, and applied Modify and Construct menu tools to turn the sketches into solids. We moved, scaled, made circular and rectangular arrays, grouped, snapped, and measured. We inserted text and placed a file into another file. In short, you learned how to use a solid modeling program!

Now join me in Chapter 4, where we'll generate construction documents of a model with the LayOut feature.

Sites to Check Out

- **Download Inkscape**   www.inkscape.org.
- **Support boards for the 123D suite**   forum.123dapp.com.
- **Thingiverse blog**   www.makerbot.com/blog/category/thingiverse/.
- **123D Gallery**   www.123dapp.com/project/search/state/all.
- **Instructables website**   www.instructables.com. Here, you can find PDFs and video instructions for lots of projects. There's a 123D group there, too.
- Follow @Autodesk123D on Twitter.
- **Creative Commons website**   http://us.creativecommons.org/. Has information about its public copyrights for intellectual property.

CHAPTER 4

Generate Construction
Documents with LayOut

MODELS ARE GREAT, but to build an item, a CNC mill or fabrication shop usually needs construction documents. These are orthographic (2D) drawings that show top, front, and side views. They're drawn to scale, meaning proportionately correct and measurable, and they are annotated, meaning they have dimensions and notes.

The good news is that you don't have to redraw the item in drafting software. A Premium feature called LayOut exports the .123dx file as a .dwg file, where it can be worked on further in AutoCAD or another .dwg-compatible program.

The .dwg file also has multiple views of the model that are visible in AutoCAD's "paper space" mode (see Figure 4-1).

If you don't have AutoCAD, you can open the .dwg in AutoCAD 360, a web app. You don't need to know how to use AutoCAD to use 360.

What Is AutoCAD 360?

AutoCAD 360 is a free viewer/editor app in which you can make rudimentary construction drawings. It lets you lightly edit and annotate

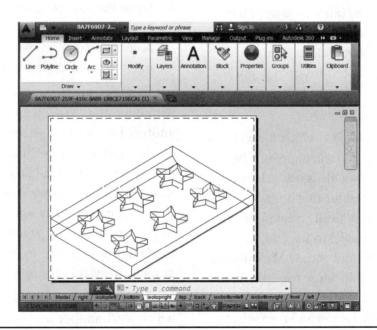

Figure 4-1   The .123dx file exported as a .dwg file, open in AutoCAD.

a drawing, create a new drawing, download a drawing in multiple formats, generate multiple 2D views of a model, as well as upload, show, and collaborate on a drawing. It has separate apps for desktop computers and mobile devices, on both Windows and Mac platforms. In this chapter, we'll use AutoCAD 360 from a Windows desktop computer to make a layout of the Jell-O tray modeled in Chapter 3.

> Don't confuse AutoCAD 360 with Autodesk 360, a subscription-based online app for editing, viewing, and sharing CAD files. Also don't confuse it with True View, a free desktop viewer for AutoCAD drawings.

Getting Started with LayOut

Let's create construction drawings from the Jell-O tray we made in Chapter 3. We'll generate 2D views from that model and annotate them.

1. Send the Jell-O tray model to LayOut by opening the 123D Design file and choosing Create 2D Layout from the 123D menu (see Figure 4-2). You'll be prompted to save the model to My Projects (also done from the 123D menu). This save must be done in the same session as LayOut (closing the model exits that session). Therefore, even if you already saved the model to My Projects in a previous session, save it again.

2. Retrieve the .dwg file. The .123dx file will be automatically converted, and you'll be emailed when the .dwg file is ready. However, you don't have to wait for that e-mail. Because the .dwg is typically created shortly after you send the model to LayOut, just go to the Sign In menu and choose My Projects. When there, hover the mouse over the small 123D mascot in the upper-right corner to access its drop-down menu and then select Messages (see Figure 4-3).

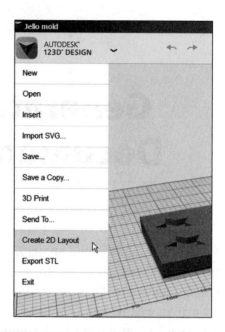

Figure 4-2   Choose Create 2D Layout from the Design menu.

3. You should see a message saying your file is now ready; inside it will be a link. Copy and paste the link into your web browser and go get your .dwg file! You'll have two options: download it directly to your computer and open it in AutoCAD 360 (see Figure 4-4).

4. Open the .dwg in AutoCAD 360. You'll need to log in to the app first. Your Autodesk account credentials will work. The app will open to reveal a plan view of the model.

Google's Chrome browser works the best with the AutoCAD 360 app, so if you're using another browser and getting script errors and crashes, just switch to Chrome. And you don't have to access Autocad 360 through an open 123D Design file. At www.autocad360.com, access it directly. After signing in, you'll be taken to your dashboard, a page with graphics of all LayOut drawings. (see Figure 4-5). Clicking a graphic opens it in the app.

The app has two screens. One is Design, in which you annotate and edit 2D views of the model, and it opens by default. The other is

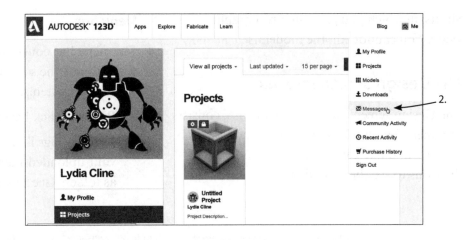

1.

Figure 4-3   Go online to My Projects and look for a message that the .dwg file is ready.

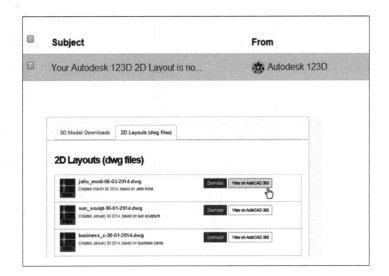

Figure 4-4   You can download the .dwg file to your computer or open it in AutoCAD 360.

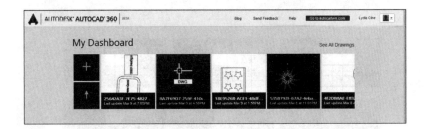

Figure 4-5   Log directly in to your AutoCAD 360 dashboard at www.autocad360.com.

Sheets, in which you generate orthographic views. You cannot edit the model itself in 360.

The Design Screen Interface

The Design screen, shown in Figure 4-6, consists of a menu bar, workspace, and palettes for tools, settings, and navigation. Run your mouse over the icons to see their tooltips.

Menu Bar

The menu bar consists of two horizontal strips at the top of the screen. The top strip contains the following items:

- **Red A**   Click here to go to your dashboard.

- **All Drawings link**   This link accesses all your uploaded LayOut files, which appear as folders (see Figure 4-7). Click the plus

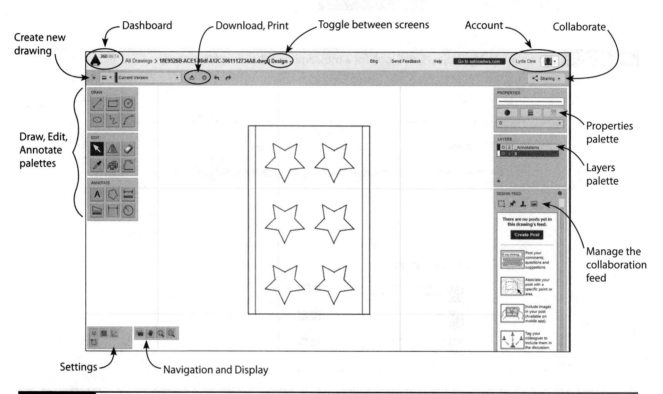

Figure 4-6   The Jell-O tray's plan view in the Design screen.

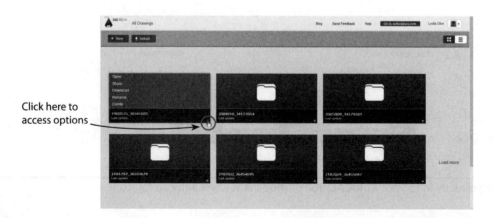

Figure 4-7   The All Drawings link leads to uploaded LayOut files and the options for them.

sign in the lower-right corner of each folder for the Open, Share, Download, Rename, and Delete options. In the page's upper-left corner are the New and Upload buttons. New opens a blank canvas to draw in, and Upload brings in a drawing from your desktop.

- **Name of drawing**   This describes the open file. Autodesk uses a lengthy string of numbers and letters, so obviously you'll want to rename the drawing (via the All Drawings link).

- **Design**   You can toggle between the Design and Sheets screens here.

- **Blog, Send Feedback, and Help**   These are links to the 360 blog, a feedback form, and a question database/message board forum.

- **Profile icon**   Used to sign out, view your account details, and connect your 360 account with other storage providers that you subscribe to (see Figure 4-8).

The menu bar's bottom strip contains the following items:

- **Plus sign (+)**   Used to create a new drawing.

- **Open Drawing**   Used to choose a different .dwg file to open.

- **Version box**   Used to view all the snapshots (versions) made of the drawing (see Figure 4-9). Snapshots are screenshots that you can take at any time, and are useful for preserving a record of alterations.

- **Download**   Used to make a .dwg, .dxf, .jpg, or .png file of the finished LayOut drawing

- **Output**   Used to print the LayOut drawing as a .pdf or .dwf.

- **Undo and Redo**   Used to undo or redo all actions, one at a time.

- **Sharing**   Used to invite collaborators and reviewers to look at the drawing.

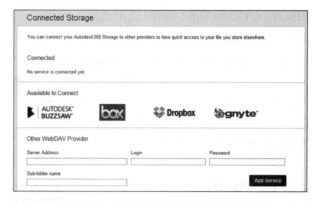

Figure 4-8   Autodesk 360 storage can be connected to other cloud storage services.

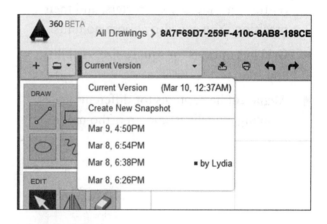

Figure 4-9   The Version box stores snapshots of the drawing.

Workspace

The workspace contains the following palettes:

- Drawing

- Editing

- Annotating

The drawing palette has tools to make a line, rectangle, circle, ellipse, sketch, and arc. The editing palette's tools let you select, mirror, trim, match properties, explode, and offset what you draw (remember, none of these apply to the model itself, which cannot be edited). The annotation palette lets you place text revision clouds, set linear and radial dimensions, and measure length and area.

To use a tool, select it and then click the screen or an item to apply it. To erase something, highlight it with the Select tool and press DELETE. Saving is done automatically as you draw, so there is no Save function.

Settings Palette

The settings palette is shown in Figure 4-10. Here, you can adjust the following controls:

- **Units** has options for inches, millimeters, centimeters, meters, kilometers, and their precision. You can also choose no units.

- **Grid** has options for square size, visibility, and snap.

- **Alignment** has options for different angles, enabling guidelines, and an ortho mode.

- **Snap** has options for the endpoint, midpoint, and center as well as for intersection and quadrants. You can also disable this feature.

Display and Navigation Palette

Here, you'll find the Pan, Zoom Extents, and Zoom Window options. There's also a setting to see and print your file as Standard (the screen's default appearance), Grayscale, or Classic (black background).

Properties Palettes

The Properties palettes are shown in Figure 4-11. You can set the line type (solid or dashed), line weight (thickness), and color as well as create layers.

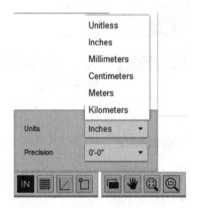

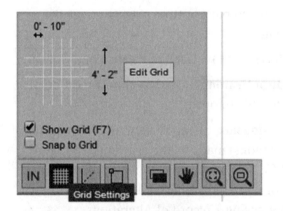

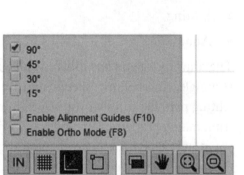

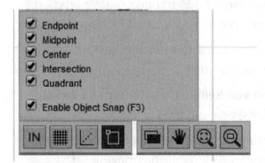

Figure 4-10 Use the settings palettes to adjust controls on the workspace.

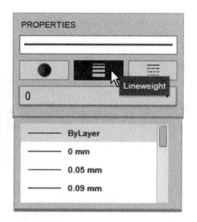

Figure 4-11 Adjust the line type, weight, and color in the Properties palettes.

Layers Palette

In the Layers palette you can make multiple layers (think of these as sheets of paper) and assign settings that affect the color, type, and thickness of lines drawn on them. You can move different features to different layers, and turn those layers on and off to print drawings showing just what you want. Adding layers is discussed later in this chapter.

Design Feed

The Design Feed is a comment stream where those individuals you invited can collaborate on the drawing. Everyone can post comments and questions, point to specific features on the drawing, and add images.

Sometimes the app can be glitchy, and the menu items described here won't show up. If that happens, first try expanding the screen. If that doesn't work, completely exit the app and then open it again. That generally fixes it.

Adding Text and Dimensions to the Drawing in the Design Screen

When you click the screen using the Text tool, a text box appears; it looks small, but will enlarge to accommodate whatever is typed inside it. When you're finished typing, click outside the box (see Figure 4-12).

Dimensions can be placed on the model or anywhere on the workspace. Click the Linear Dimension tool on one corner of the model, then on a second, and then move the dimension line where wanted. Click again to finish (see

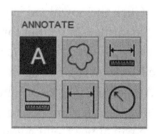

Figure 4-12 Add notes with the Text tool. Click it in two places to make a text box.

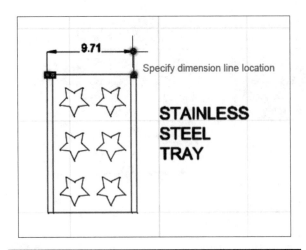

Figure 4-13 This dimension note was made with the Linear Dimension tool.

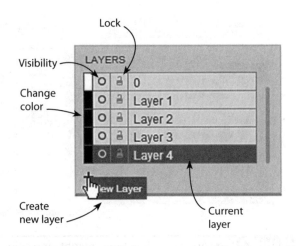

Figure 4-14 Click the plus sign to add a new layer.

Figure 4-13). Press ENTER to immediately return to the tool.

You can't make leader lines (lines with arrowheads that point to features), nor can you adjust characteristics or font size. However, dimension units can be changed with the Units setting, and dimension and text color can be changed with the Properties palettes.

Adding Layers in the Design Screen

Layers are sheets that can be turned on and off, enabling you to display or print precisely what you want. Click the plus sign in the Layers palette to make new layers (see Figure 4-14). Select a layer (it will highlight blue) to make it active; whatever you draw goes on the active layer. You cannot move items to different layers later. Click the oval to turn a layer off, making everything on it invisible. Click the lock to make everything on the layer unalterable. Right-click a layer to rename or delete it.

Click the black rectangle to open a color palette, and then click a color to assign to that layer. Everything you draw on it will be that color (see Figure 4-15). To change an item's color later, change the color of the layer it's on.

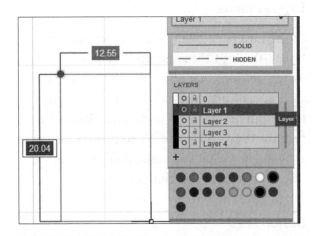

Figure 4-15 Assign a color to a layer, and everything drawn on it will be that color.

Downloading and Printing in the Design Screen

When you're done with the drawing, you can either download or output it:

- **Download** Click the download icon and choose the file format. Note that for .dwg and .dxf files, you'll need to choose a specific version (see Figure 4-16). Saving in an older version of AutoCAD enables collaborators who don't have the latest version to open it. A message will appear telling you the

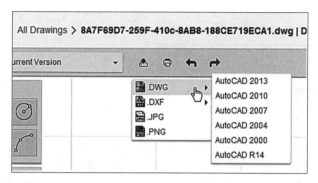

Figure 4-16 The drawing can be downloaded in multiple file formats.

file is downloading. When the download is finished, look for the file in the downloads folder or wherever your downloads appear. All changes made in the AutoCAD 360 app will appear in the new file (see Figure 4-17).

■ **Output**   Click the print icon and choose the output—that is, its format (PDF or DWF) display, (paper) size, and orientation (see Figure 4-18). You can't choose a scale here; that's done on the Sheets screen. A pop-up with a download link will appear when the file is done.

The Sheets Screen Interface

Toggle to the Sheets screen, as shown in Figure 4-19. Like the Design screen, it has a menu bar and workspace (see Figure 4-20). Note the navigation panel at the bottom, from which you can pan, zoom, and adjust the display setting. You can also create a new drawing, access other drawings from the dashboard, and collaborate.

On the workspace's right side is a pane with graphics of different orthographic and isometric views (see Figure 4-21). Click each graphic to generate that view. You can then download or print it, including printing to scale.

You can't annotate in this screen, and you can't generate alternate views in the Design screen. Hence, dimensions and notes can only

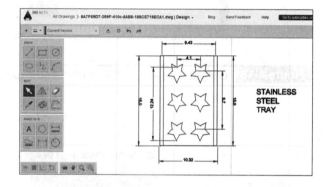

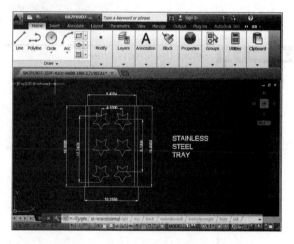

Figure 4-17   The AutoCAD 360 drawing and the downloaded .dwg file open in AutoCAD.

be added to the plan view. To annotate all views and do more work on this file, import it into AutoCAD or other .dwg-compatible software.

You can download or output the view by clicking the appropriate icon in the menu bar.

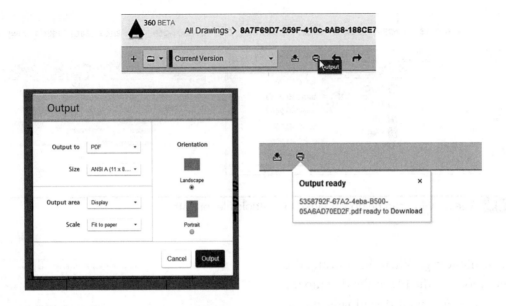

Figure 4-18   Output the file to a PDF or DWF and choose its options.

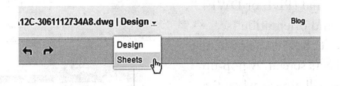

Figure 4-19   Select the Sheets screen.

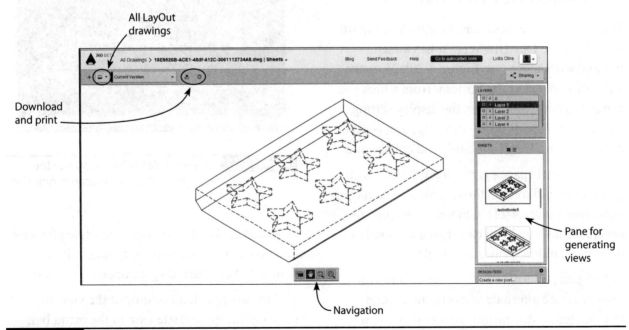

Figure 4-20   The Sheets Screen interface.

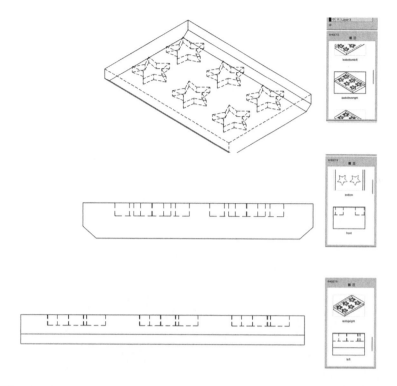

Figure 4-21 Click a graphic to generate a view.

Note that when outputting, you can select a scale (see Figure 4-22).

Summary

In this chapter, we exported a 123D Design model into a .dwg file. Then we opened it in a web app called AutoCAD 360, where we generated orthographic views and annotated one of those views. Such drawings are useful when the item will be made using a CNC machine or other fabrication process.

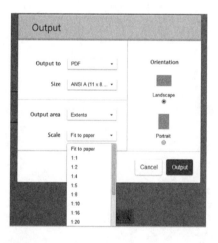

Figure 4-22 You can scale the drawing in the Sheets screen.

Sites to Check Out

■ For a free 30-day full version or three-year student AutoCAD trial, go to www .autodesk.com/products/autodesk-autocad /free-trial.

■ The Autocad 360 support forum is at https:// getsatisfaction.com/autocad360.

■ Follow @Autocad360 on Twitter. The team is also on Facebook.

CHAPTER 5

Capture It! with 123D Catch

MODELING ISN'T LIMITED TO conventional programs with drawing and editing tools. Modeling can also be done with photographs. In this chapter our tools will be a camera and on-line *reality capture* software. The result is called a "capture" or "photoscene." What it is, is a *photorealistic* or *phototextured* model, a mesh model covered with photos to make it closely resemble the subject. If you want to make a model that's easily recognized as a specific person or object, as shown in Figure 5-1, Autodesk's 123D Catch apps are for you.

The technical name for reality capture is *photogrammetry,* a technique that takes multiple, overlapping photos of a subject and uploads them to a server. Processing software stitches them together by finding and matching the photos' common features and calculating spatial distances between them. The result is a colored, textured mesh model that you can download as a .3dp or .obj file (the latter being 3D-printable). Think of this as 3D scanning.

Reality capture is the most efficient way to model whole environments, such as a room, backyard, or city. It enables you to start with an existing product instead of a blank slate, making the design process faster and more efficient. Professional-level software such as Autodesk Recap is used in diverse fields. Real estate agents capture homes for client presentations. Archeologists duplicate artifacts by capturing them, which avoids the damage caused by physical molds. Museums upload captures of

Figure 5-1 A capture of a sculpture at the Asian Art Museum in San Francisco. Courtesy Christian Pramuk, Autodesk.

their collections for the online world to admire. Town planners capture whole neighborhoods to study. Cartographers capture terrain to make maps. Such projects may use sophisticated methods such as attaching a GoPro camera to a quadcopter (drone) to take thousands of photos. Entrepreneurs have used reality capture software to produce whimsical products such as a chocolate mold of a head, a family's faces on chess pieces, a pet's likeness on a candy tablet dispenser, the bride and groom on a cake topper, and assets for video games. All of these tasks would be difficult or impossible to do with conventional modeling programs.

What Is 123D Catch?

123D Catch is an entry-level reality capture program that lets you explore the potential of this modeling method. It has three steps: capture, process, and share. *Capture* is photo-taking, and is done with a smartphone, iPad, or digital camera. *Process* is the online stitching into a photo scene. An Internet connection is required to upload the photos (collectively called the photoset), but the processing is done on Autodesk's servers. *Share* is sending the finished file to My Models, where you can post it in the 123D Gallery or keep it private. You also download the .3dp and .obj files from there.

You can access Catch at www.123dapp.com/catch. It has three platforms: one for the iPhone, iPad, or Android devices (with links to the App Store or Google Play); one for the web; and one for a PC.

- **iPhone/iPad/Android app** This app captures the subject with the phone or tablet's built-in camera, and directly uploads the photoset to Autodesk's servers for processing. When the process is finished, you have the option to save the file to My Models or delete it. This app has no tools for working on a file.

- **Web app** This app makes captures by uploading photos from your computer to the cloud, or 123D Catch servers. It can import small Catch .3dp files and has tools for making simple edits and repairs. Use Google's Chrome browser to run the web app because other browsers either won't work or won't work optimally (some tools may not function properly). Chrome will prompt you to download a plugin for repair functionality.

- **Desktop app** This app is installed on your computer. It makes captures by uploading

stored photos to Autodesk's servers. It can import larger Catch .3dp files than the web app can. It has different tools than the web app, too; it can scale the capture to its actual dimensions (captures have no scale when made), change its mesh density, manually stitch more photos in, create an animation file, and more. You can bounce a small file between the web and desktop apps to utilize both their features.

We'll explore these apps in the following sections.

Catch on the iPhone/iPad

 To begin, launch the app. We'll use an iPad to make the capture for this example.

You don't have to sign in yet, but if you do, the middle part of Figure 5-2 shows what you'll see. It's your feed, and the models of Catchers you follow will appear there. When you're not logged in, featured models appear. Tap the bars on the left to open the screen shown on the left. Read its options. The Home option takes you to your 123D account online. Tap the plus sign on the right to open the screen on the right. The large white plus sign starts a new capture. Tap it to do just that. Your iPad screen will look like Figure 5-3.

The large white circle is the camera button. The wheel in the lower-left corner shows your movement around the subject, which can help you maintain a consistent distance between shots. Toggle it off, if you want, by tapping the smaller wheel. The check mark at the top finishes the catch.

Our subject for this example is the little cottage shown in Figure 5-4. It's on a table whose location enables movement in a complete circle around it. The surface under the cottage is gift paper that was selected for its color/pattern contrast with the cottage. Figure 5-5 shows it as seen through the iPad.

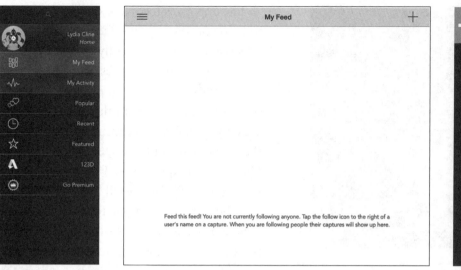

Figure 5-2 The iPad screen when Catch is launched and you're logged in to your 123D account. Tap the bars to open the left screen; tap the plus sign to open the right screen. Tap the large white plus sign to start a capture.

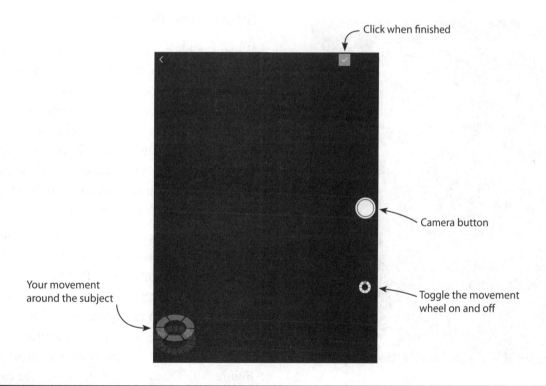

Click when finished

Camera button

Your movement around the subject

Toggle the movement wheel on and off

Figure 5-3 The Catch screen

Figure 5-4 This cottage is the subject of our capture. The paper under it was chosen for its contrast with the cottage's colors.

Figure 5-5 The cottage seen though an iPad.

The app gives feedback during the process to help you shoot good photos. For instance, it tells you if a photo is badly exposed and asks if you want to delete or reshoot. Reshooting superimposes the existing photo on the frame so you can reshoot from the same location. When you're finished capturing the subject, tap the check mark in the upper-right corner. The photoset will appear (see Figure 5-6). Review the thumbnails for overall lighting consistency, and then review each one for focus and lighting. Delete or reshoot any photos that are blurry, too dark, or too light. Also delete ones with large glare spots, even if the spots are in the background. Keep the number of photos to 70 or fewer. Tips for photographing the subject are at the end of this chapter.

Submit the Photoset

When the photoset is in order, tap Submit in the upper-right corner. You'll be prompted to sign in to your Autodesk account, if you aren't already. The app will search for a Wi-Fi connection. If it can't find one, it won't use your data plan until you tap the Go Ahead button. If you're worried about inadvertently using the data plan, you can disable it in the iPad's settings menu (see Figure 5-7).

Processing generally takes less than 30 minutes, depending on how busy the servers are and where in line your capture is. The iPad's inactivity lock mode doesn't affect transmission. You'll get status updates, as shown in Figure 5-8, and a message will appear on the app (and on the other Catch apps as well) when the processing is finished (see Figure 5-9). Open the message to see the capture. It will display in a 3D viewer and may be upside down, at a weird angle, and with bits (or lots) of background floating around it. That background can be removed in Catch's web or desktop app. Straighten out the capture by pinching and dragging it. Ideally, it will be clean, hollow, complete on all sides, and look like the subject.

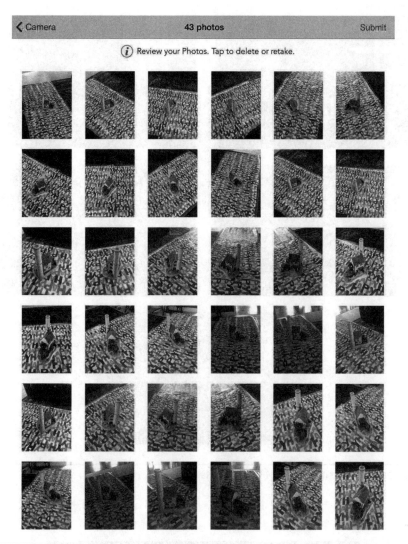

Figure 5-6 The photoset appears after you click the check mark. Review it for lighting and focus.

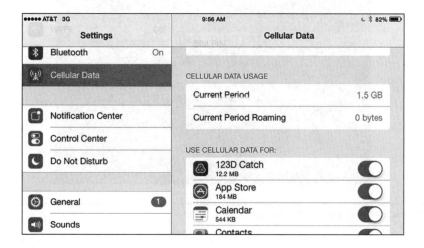

Figure 5-7 Catch's access to your data plan can be disabled in the iPad's settings.

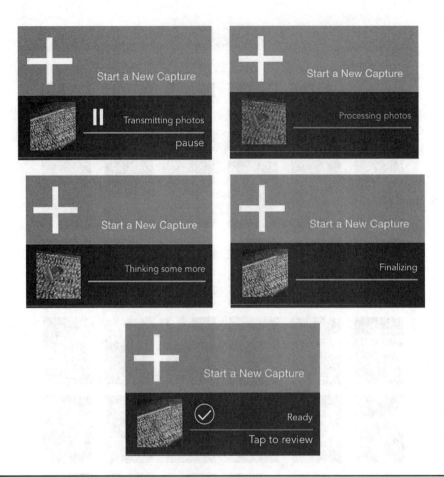

Figure 5-8   Updates keep you informed of the processing status.

Figure 5-9   A notice appears when the capture is finished.

At the bottom of the screen is a graphic for messaging or e-mailing a link to the capture, as shown in Figure 5-10, and a dialog box asking if you want to keep or discard it. If you tap "keep" (the check mark), you'll be asked to frame the capture (pinch and drag until it looks suitable for display). Tap Publish, and off it goes to your 123D account (see Figure 5-11). When your capture gets there, you'll be asked to name, describe, and tag it (see Figure 5-12). You can also e-mail yourself a link to it. Finally, tap Share. Despite its name, Share just saves the capture to My Models, where you can opt to display it in the 123D Gallery or keep it private.

Know that you can go to the feed screen at any time. Tap the bars to access the left menu,

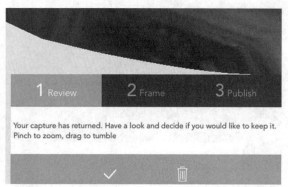

Figure 5-10 You can e-mail or message a link to the capture here, as well as keep or discard it.

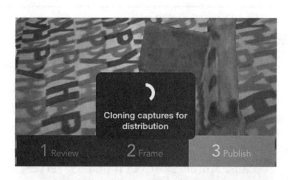

Figure 5-11 Tap "Publish" to send the capture to My Models in your 123D account.

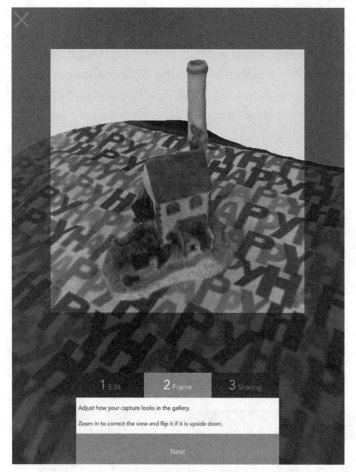

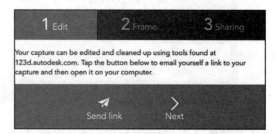

1. Adjust how the capture will look in the gallery

2. Name and describe it

3. E-mail yourself a link to the capture

Figure 5-12   Name, label, and share the model. E-mail yourself a link to it, if you wish.

tap Home to view your saved capture, and then tap the capture (or any other one there) to load it (see Figure 5-13).

Download the Capture

To download the capture, go to 123Dapp.com, sign in to your account, and click My Models. Find the model, click the Edit/Download button (see Figure 5-14), and select Download 3D Models. You'll get a ZIP file containing the model in .obj and .3dp formats (see Figure 5-15a). Extract the contents and then rename the .obj and

.3dp filenames with more descriptive ones. You'll see some other files in the folder; keep everything because Catch references these files when opening the capture. The photoset doesn't download, because Autodesk stores it on their servers and Catch references it there. Sharing the .3dp file with others requires them to have Catch installed, too, and their app will reference the stored photoset.

Check "Photo Package" to see the downloaded photos. Sometimes the file extension gets stripped off during the download (a glitch), and the JPGs will look like Figure 5-15b. If you want to restore

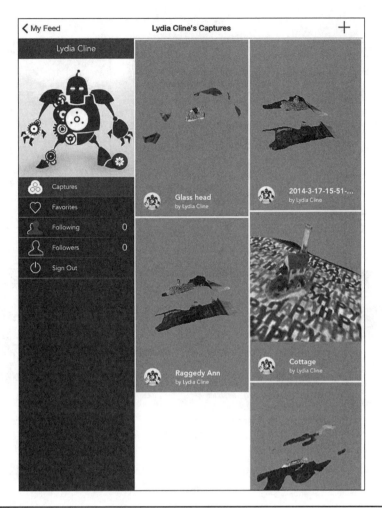

Figure 5-13 Access all captures through the left menu bar.

Figure 5-14 At My Models, click the model and then click Edit/Download.

(a)

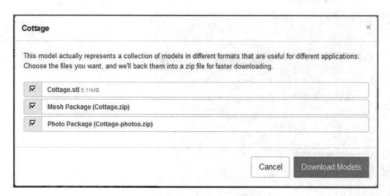 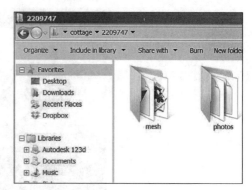

1. Download the capture   2. Find the download file

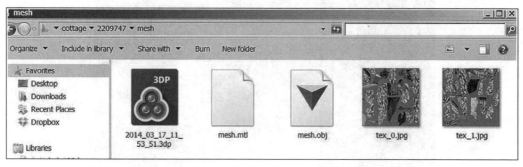

3. Find the 3DP file inside the folder

(b)

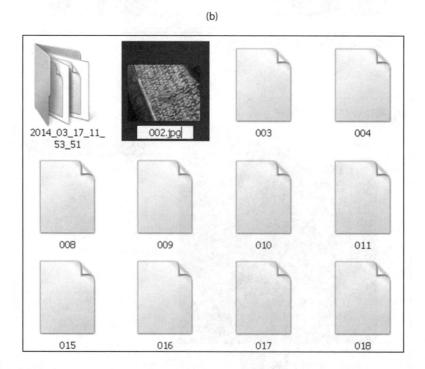

Figure 5-15   (a) The model downloads as a ZIP package containing .stl and .3dp files, which you can find in the "mesh" folder. (b) The obj folder contains the JPG files, but you may have to restore the extension.

the extension so that you can open a file, click the filename, type **.jpg** at the end (include the period), and the file will appear correctly.

Troubleshoot the iPhone/iPad App

You may encounter some problems with the mobile app. A common one is the photoset getting stuck in transmission. If over two hours have passed and the capture is still uploading, it will probably never upload, even if it still says "transmitting." Here are some suggestions:

- Tap the Submit button again.
- Do a small catch of 10 photos and submit them. This might prompt the first one to resume uploading.
- Log out and then back in to your 123D account.
- Do a hard reboot of the iPad/iPhone.
- Restart or reinstall the app.
- If you're on cellular data, switch to Wi-Fi.
- Upload the photoset through the web or desktop app.
- If you get an error message when trying to download the capture from My Models, launch the Catch web app and download the model from inside it.

If you have to reboot, restart, or reinstall, you can find your images in your device's My Photo Stream. You won't be able to resubmit them through the mobile app, but you can do so through the web or desktop app.

The Catch Web App

If you need to clean, repair, or change the model, you can so in Catch's web or desktop app, in Meshmixer, or in another modeling program such as Autodesk Mudbox, Maya, or 3ds Max.

To access the Catch web app, point the Google Chrome browser to www.123dapp.com /catch and click Launch 123D Catch online (see Figure 5-16a). A splash screen will appear (see Figure 5-16b); if you don't see everything shown in Figure 5-16b, maximize the screen. You can make a new capture by clicking the Start a New Project button to upload photos from your computer. However, we're not making a new capture, so just click the X in the splash screen's upper-right corner to expose the web app behind it (see Figure 5-16c).

The Web App Interface

Figure 5-17 shows the workspace in the web app's interface. It contains Sign In and Help menus as well as a menu bar, work canvas, and navigation panel.

Sign In and Help

The Sign In and Help menus appear in the upper-right corner (see Figure 5-18). Signing in accesses your 123D account. You need to be online to utilize Help because there is no local Help.

Menu Bar

The menu bar contains the Utilities menu (see Figure 5-19) and tools. The red dot means that new captures await in My Projects. Click the bars to open the utility menu. It contains the following items:

- **New**   This item creates a new model (the same thing you could have done from the splash screen).

(a) The home page

(b) The splash screen

Close the splash screen to expose the app behind it.

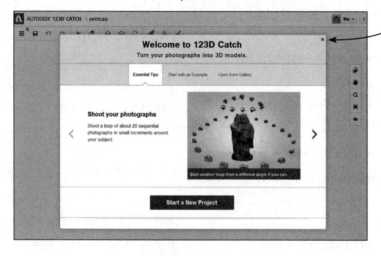

(c) The web app

Figure 5-16 With Catch's Web app you can submit a new capture or work on an existing one.

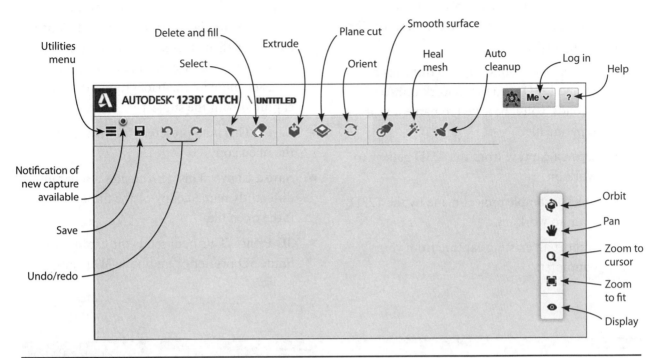

Figure 5-17 The Catch Web app interface

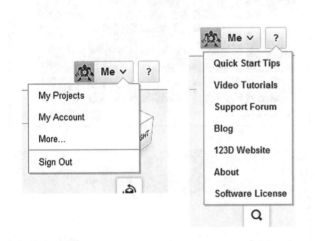

Figure 5-18 The Sign In and Help menus

Figure 5-19 The Utilities menu

- **Open**   This item brings up a browser (see Figure 5-20) from which you can perform the following tasks:

  - Import an existing capture from your online 123D account to work on. After it loads, you'll be asked to name it as a separate file.

  - Open a capture from the 123D gallery to work on.

  - Open a sample project made by the 123D team to work on.

  - Import an existing capture from your computer.

- **Edit**   This item lets you add or subtract pictures from the capture's photoset. A new capture is then created, which you'll name separately, as shown in Figure 5-21. As an aside, note all the background that got captured with the subject.

- **Save**   This item preserves all work done on the open copy.

- **Save a Copy**   This item creates a separate, differently named copy of the file, and makes it the open file.

- **3D Print**   This item sends the capture to a home 3D printer or an online 3D printing service.

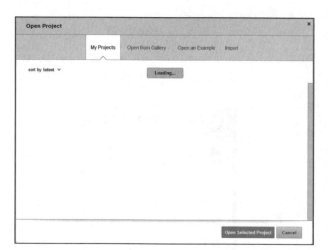

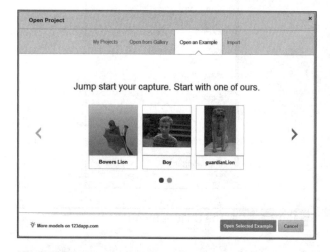

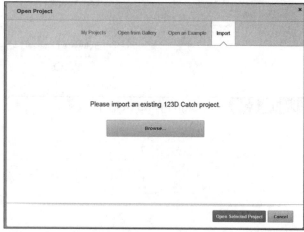

Figure 5-20   Options on the Open submenu

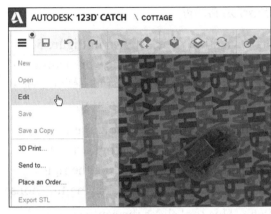

1.

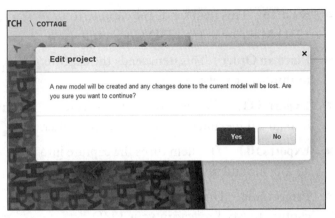

2.

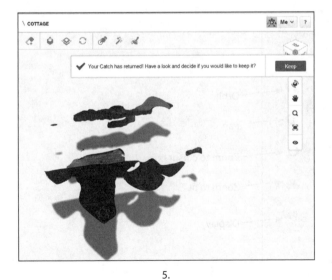

3.

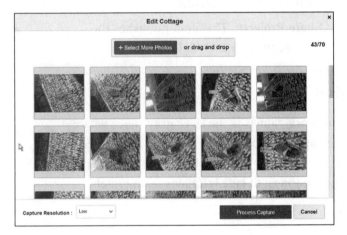

4.

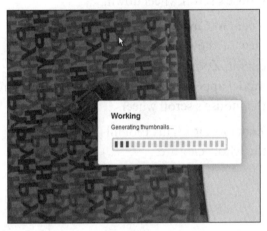

5.

6.

Figure 5-21 Click Edit on the Utilities menu to add or subtract photos to/from an existing capture. This results in a new, separately named capture.

- **Send To**   This item sends the capture to 123D Make or to the 123D CNC utility.

- **Place an Order**   This item sends the capture to third-party service.

- **Export STL**   This item turns the capture into an .stl file and saves it on your computer.

- **Export OBJ**   This item turns the capture into an .obj file and saves it on your computer.

- **Project Info**   This item lets you save the capture to My Projects in your 123D account, where you'll describe it, tag it, and set it as public or private.

- **Share Project**   This item publishes the capture to the 123D Gallery.

Work Canvas

The work canvas is the large gray space where the capture appears and is viewed and edited. At its top is a tool bar, which we'll discuss shortly, and on the right side is a navigation panel.

Navigation Panel

The navigation panel, shown in Figure 5-22, appears on the right side of the work canvas. It contains the following tools, which let you view and display the model in different ways:

- **Orbit**   This tool rotates you, the viewer, around the capture at all angles and heights. You can also orbit by holding the right mouse button down and dragging the mouse.

- **Pan**   This tool slides the capture around the workspace. You can also pan by holding the mouse's scroll wheel down.

- **Zoom to Cursor**   This tool enlarges your view of the model and centers it on the cursor. Zoom in and out (that is, make the model appear larger and smaller) by rotating the mouse's scroll wheel.

- **Zoom to Fit**   This tool fills the screen with the whole capture.

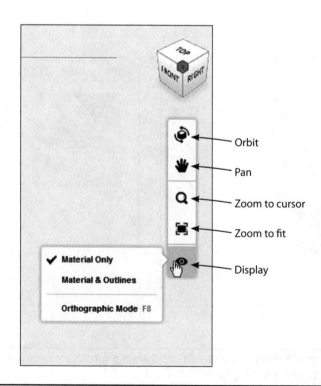

Figure 5-22   The navigation panel, with the Display icon's options shown

■ **Display** This tool lets you view the capture as "material only" or "material and outlines." You can also view it in Orthographic mode, meaning parallel lines don't converge like in the default perspective mode.

When a capture is loaded, a ViewCube appears (the box on top of the panel). This shows the capture's orientation on the work plane. Click the left mouse button on the ViewCube and drag to rotate it; you'll see that the capture rotates as well. Clicking the ViewCube's planes displays the capture from the six sides (top, bottom, front, back, left, and right). These are perspective views, though (that is, they're 3D drawings with line convergence).

Set the Display icon to Orthographic Mode to make these views true orthographic ones (2D drawings with no line convergence). This setting will remain until you change it back.

Import the Cottage .3dp File

Let's import the .3dp file of the cottage from our computer and work on it. At the Utility menu, click Open/Import and then navigate to the .3dp file, as shown in Figure 5-23a.

I've temporarily chosen the Materials and Outlines option to show the mesh (see Figure 5-23b). There are three densities, which we'll talk about later; this is the middle one, called

(a)

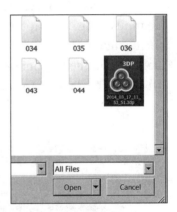

(b)

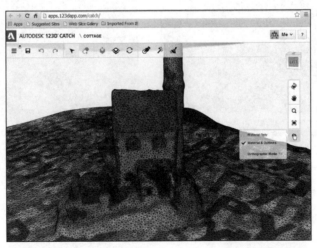

Figure 5-23 (a) Navigate to the .3dp file of the cottage that was downloaded to the computer, and import it into the web app. (b) The capture displayed with the Materials and Outlines option, showing its standard density mesh.

"standard." It's likely there will be numerous holes in the mesh that will need repairing before the capture can be 3D-printed.

Edit and Repair with the Web App

In order to edit and repair the capture, we need to know what the menu bar's tools can do. Hover over each one to see an info tip describing it. When a tool is selected, it turns black and a bar with relevant adjustment options appears at the bottom of the screen.

The Select Tool

 This tool highlights the mesh for editing. It can be applied as a paint spot or lasso (see Figure 5-24). Click the mouse onto the capture to paint-select; click off the model to lasso-select. You can also move the slider on the bottom bar all the way to the left to make a lasso. Moving it to the right enlarges the paint spot. To deselect a spot, hold the SHIFT key down and click the Select tool on it. Press ESC to deselect everything.

Everything inside the lasso or paint spot will be edited. Delete the selection by clicking the Delete button on the adjustments bar. Click the Invert icon to select everything outside the lasso or paint spot. This enables you to quickly remove most extraneous material. The Delete button doesn't appear in Orbit mode, only in Select mode (this is when the arrow icon is highlighted). Sometimes a selection takes some time to apply (maybe five or ten seconds).

You can't perform fine selecting in this app. Zooming in closely to select background often results in part of the capture getting selected with it (see Figure 5-25). The paint spot finesses better than the lasso, but it, too, may select more than you want. The Smooth Selection Boundary icon next to the Invert Selection icon smooths the boundary of the selection.

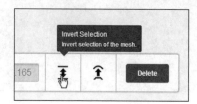

Figure 5-24 With everything inside the lasso or paint spot selected, click the Invert Selection icon to select everything outside the lasso or paint spot.

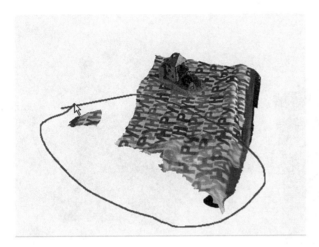

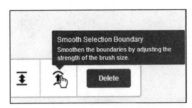

Figure 5-25	After you remove most of the surface or background, fine-tuning with the Select tool becomes difficult, due to it selecting more than you may have wanted.

The Delete and Fill Tool

 This tool deletes a selected part of the mesh and then fills it with a generic, smoothed mesh. This is useful to remove, or "brush away," an unwanted part of the catch.

The Extrude Tool

 This tool gives volume to a selected surface (see Figure 5-26). Select the surface and then click the tool. The extrusion will automatically occur. Adjust the extrusion length with the slider or by typing a number. Click the direction icon in the bar for options to extrude along a specific axis.

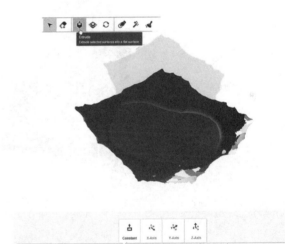

Figure 5-26	The Extrude tool pulls a selected surface out.

The Plane Cut Tool

 This tool cuts through the whole capture at a location of your choice and removes a portion. When you

click the Plane Cut tool onto the capture, the cutting plane, an arrow, and round button appear (see Figure 5-27). Drag the arrow up and down to move the plane up and down. Drag the button to rotate the plane. You can orient the

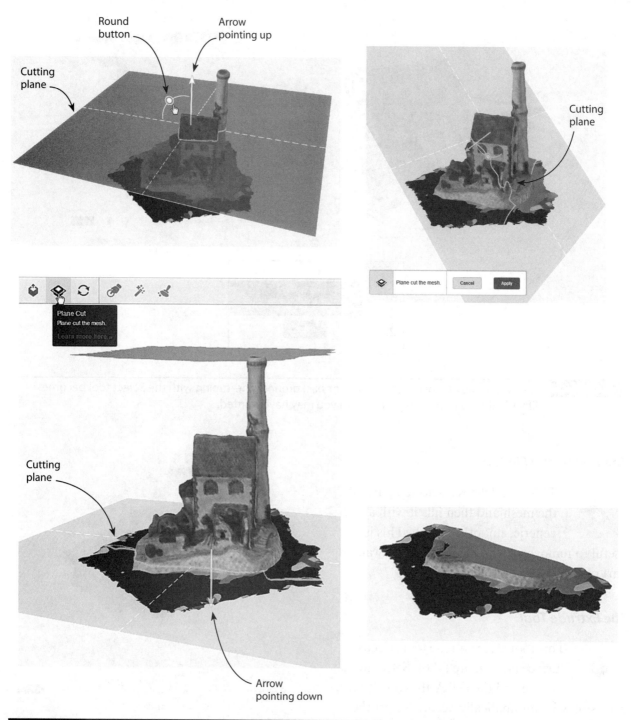

Figure 5-27   The Plane Cut tool cuts through the capture and removes the portion opposite the arrowhead.

plane horizontally, vertically, or any angle in between. When the arrow points up, everything below the cutting plane will be removed. When the arrow points down, everything above the cutting plane will be removed. When the arrow is angled, the cutting plane follows the contours of the capture. The spot above the cottage in the third graphic is a glitchy shadow (it's supposed to define the floor of the model space, but is inverted).

The Orient Tool

 This tool rotates the capture. One of Catch's built-in examples is demonstrated in Figure 5-28. Click the Orient tool. The whole capture selects (note the thick line outline), and a plane and handle appear. Click the mouse on the handle to select it (it will turn yellow). Then use that handle to rotate the model. Click anywhere on the work canvas to set the new position.

The Smooth Surface Tool

 Select a size for this tool and then slide it over a ripply mesh to smooth it (see Figure 5-29). This is useful when you want to extrude that portion of the mesh to create a smooth pedestal base.

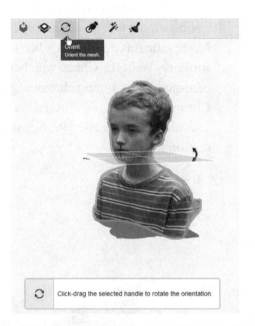
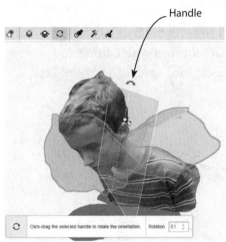

Figure 5-28 The Orient tool rotates the model.

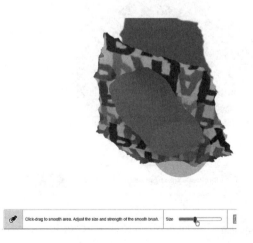

Figure 5-29 The ripply mesh on the left was smoothed by rubbing the Smooth Surface tool over it.

The Heal Mesh Tool

 Select the whole capture and then click this tool onto it. Heal Mesh will automatically searches for holes. Pins appear at small holes so you can heal them individually (see Figure 5-30a). You can also fill all holes at once, as the second graphic in Figure 5-30b shows. Remember to select the capture before selecting the tool.

The Auto Clean-Up Tool

 Select the whole capture or a portion of it and then click this tool. It removes disconnected vertices and edges of the mesh that are in the selection.

When you are finished with editing and repairing your capture, save it to My Projects.

Troubleshoot the Web App

Being new and constantly under development, the web app can be glitchy. You may have problems loading it or importing the capture. Here are some common problems and possible solutions:

Problem:

- The web app won't load.

Possible solutions:

- There are many hardware, browser, and operating system combinations; yours may not be supported. Catch uses Windows and Mac platforms and needs a browser that supports WebGL. Check whether WebGL is turned on by typing **chrome://gpu/** in your Chrome browser. If WebGL is reported as unavailable, check the "Problems Detected" section for why.

(a)

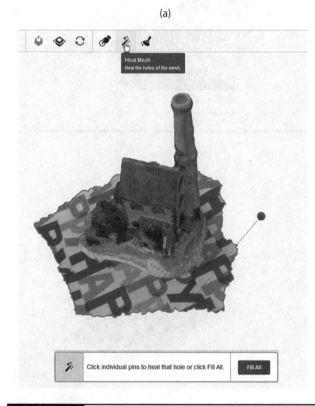

(b)

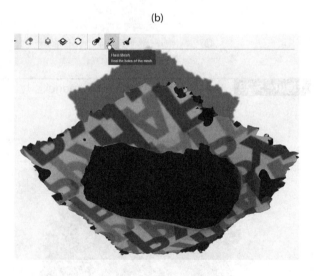

Figure 5-30   The Heal Mesh tool searches for holes and lets you fix them individually or all at once.

- The graphics card driver and operating system are not up to date. When you update the operating system, update the app (reinstall it).

- Catch uses the Chrome, Safari, and Firefox browsers, but Chrome works the best. However, when it (or another browser) is updated, sometimes a piece of code that's needed to run the app gets inadvertently changed or deleted. Try a different browser.

Problem:

- The capture doesn't load (see Figure 5-31).

- The capture loads, but crashes shortly after.

- The plugin crashes.

- The page becomes unresponsive.

Possible solutions:

- Work on the capture in 123D Meshmixer instead.

- The web app works well on small files ("small" being arbitrary, depending on the number of photos, resolution, and even the subject). Remove photos from the photoset to make the capture smaller. However, know that this does affect its quality.

- Open one of the sample captures (Utility menu | Open | Open an Example), and practice on it if your goal is just to learn how the app works.

The Catch Desktop App

The desktop app's functions are different from the web app's functions. It is also more stable and can handle larger files. That said, be aware that like the other apps, it iterates frequently, glitches are fixed as they're brought to the developers' attention, and some legacy functions hang around from the original Photofly beta

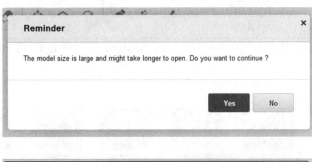

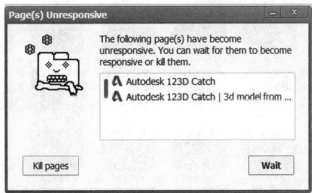

Figure 5-31   Error messages that may appear.

even though they no longer work (such as "splats" in the View/Display menu).

We'll make a new capture with this app, delete the background, scale it, stitch photos change the origin, make an animation, and export the capture into a different format. To begin,

 download the desktop app at www.123dapp.com/catch. Install and click the icon to launch the app.

The subject of our new capture is the glass mannequin head shown in Figure 5-32. Glass must be covered with an opaque material in order to capture it. I dampened the head and patted flour all over it. Then I took 40 photos with a digital camera (see Figure 5-33) and copied them to a desktop computer.

Click Get Started. You'll be prompted to log into your 123D account. Log in and then

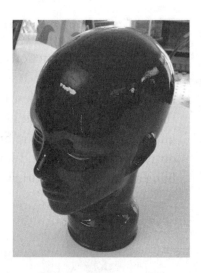 

Figure 5-32   On the left is the glass mannequin head; on the right, it's covered in flour to make it opaque enough to capture.

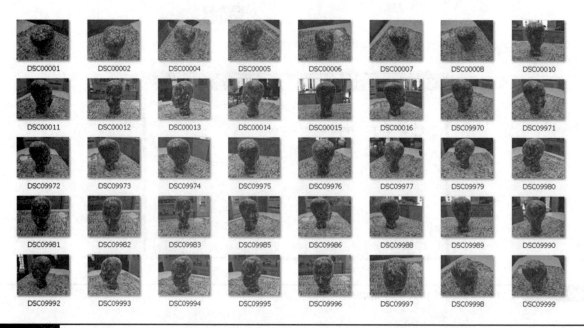

Figure 5-33   The photoset of the glass head

navigate to the photoset. Drag a window around the photos to select them, and click Open (see Figure 5-34).

Name and describe the capture. You'll be asked if you want to wait for it or be e-mailed when it's ready. If you choose e-mail, the app will quit and you'll be sent a download link to the .3dp file when it's finished. If you choose to wait, you'll stay online. The capture will appear in a viewer when finished, and the .3dp file will be saved in the same folder as the photoset. Make your choice, click the green check mark, agree to the terms of service, and you're off (see Figure 5-35). If you want an .stl file of it, go to your online 123D account, click My Models and find the capture. Open it and click the Edit/Download button to download an .stl, as well as .obj and .3dp files.

Click Get Started

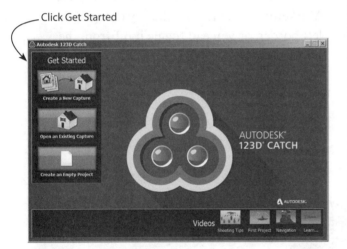
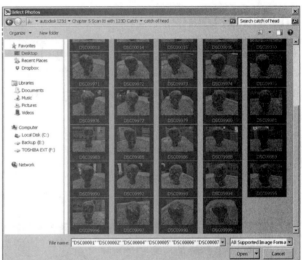

Figure 5-34 Launch Catch, click Get Started, browse to the glass head photoset, and select and import it.

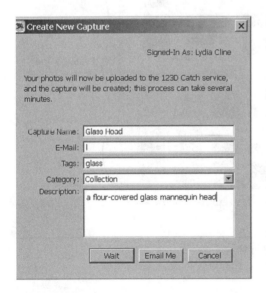

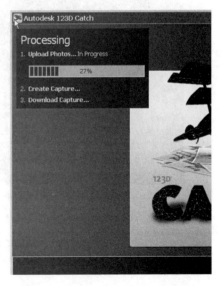

Figure 5-35 Name and describe the glass head capture. Then click the green check mark to start processing it.

Preserve a local copy of the .3dp file because there are no undo and redo functions in the desktop app. Then click it to open it in the Catch desktop app (see Figure 5-36). The next time you launch the app, recently opened files will appear as thumbnails at the bottom of the screen. You can click them to quickly open the file.

Note that there are two windows. The left is the interface. The right is the Marketplace, a collection of online materials.

The Marketplace

The Marketplace shows featured and popular captures. Clicking one closes the current file in the interface and loads the one you clicked.

At the top of the screen is My Projects, where you can access your 123D account. Click to see thumbnails of all your captures; again, clicking one closes the current file and opens the one you clicked. However, you can run multiple instances of the app, if you want, by clicking the Catch icon on your computer's desktop. If you have difficulty loading one of your captures from the desktop, try loading it from My Projects.

The Publish to the Gallery button uploads and publishes the current file after asking you to describe and tag it. You can make the Marketplace window smaller by dragging its left border, or you can toggle it off from the interface window (File | Marketplace | Toggle).

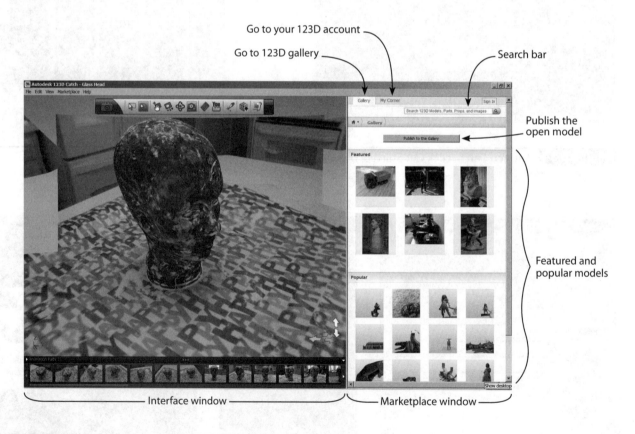

Figure 5-36   The glass head capture loaded into the Catch desktop app

The Desktop App Interface

The desktop app's interface has a work canvas, menu bar, toolbar, and World Coordinate System (WCS) icon (see Figure 5-37). There's a lot here; we'll look at the most commonly used features and do some small projects with them.

Work Canvas

The work canvas is the gray space the model occupies. The photoset strip (photos from which the photoscene was made) appears at the bottom of the canvas.

Menu Bar

The menu bar appears at the top of the work canvas. Its categories are as follows:

■ **File**   This item includes standard utilities, such as Open, Save, Export, Quit, and 3D Print. There's also a Preferences submenu that has settings and shortcut options. One option is the ability to change Catch's navigation system to match that of some popular Autodesk programs (see Figure 5-38).

■ **Edit**   This item includes tools for selecting, scaling, changing the origin, and adding more photos to the capture.

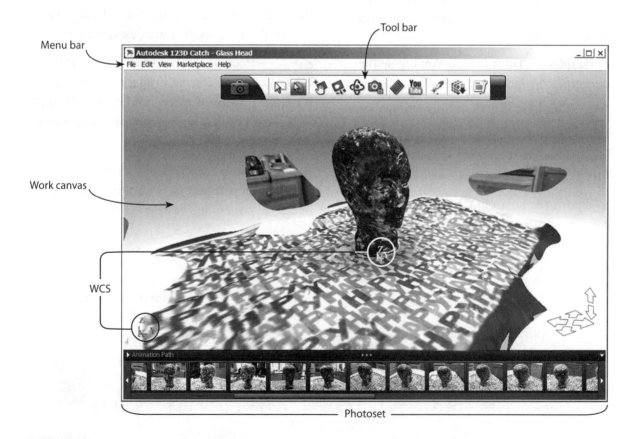

Figure 5-37   The desktop app's interface

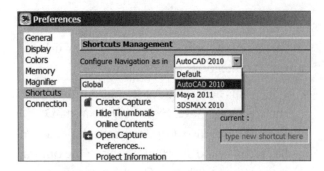

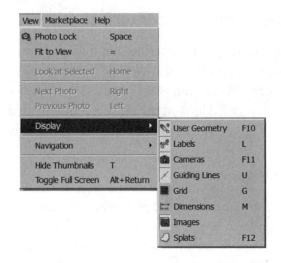

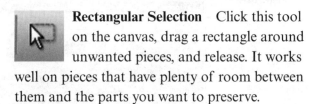

Figure 5-38 By selecting File | Preferences | Shortcuts, you can change Catch's navigation system to match that of some other Autodesk programs.

Figure 5-39 The View menu contains display settings and navigation tools.

- **View** This item includes display settings and navigation tools (see Figure 5-39). Display settings control features visible on the screen, such as cameras, dimensions, and gridlines. Navigation tools move you around the work canvas.

- **Marketplace** This item lets you toggle the Marketplace window on and off.

- **Help** This item includes information about your capture and a link to the Catch website.

Toolbar

The toolbar is the horizontal bar at the top of the work canvas (see Figure 5-40). Relocate it by dragging its ends. Activate its tools by clicking them; an activated tool has a blue background. The tools in the toolbar are described next:

Rectangular Selection Click this tool on the canvas, drag a rectangle around unwanted pieces, and release. It works well on pieces that have plenty of room between them and the parts you want to preserve.

Lasso Selection Click this tool on the canvas, drag a lasso around unwanted pieces, and release (see Figure 5-41).

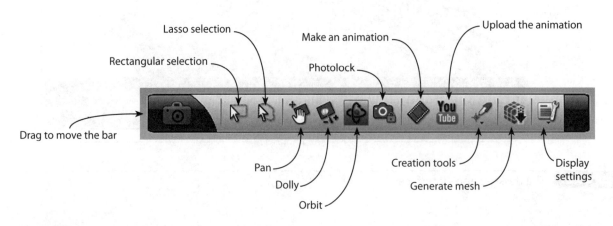

Figure 5-40 The toolbar

This works well on pieces that are close to parts you want to keep. The lasso's dashed line is a selection boundary.

Hold the SHIFT key down to select multiple pieces in one operation. Then delete them by pressing the DELETE key or by going to the Edit menu and choosing Delete. Deselect by clicking twice or pressing the ESC key. Using these tools is easiest when you're in 3D navigation mode (discussed shortly), and the cameras are turned off (select View menu | Display | Cameras). As mentioned earlier, there are no undo and redo functions, so use the Edit/Save As option frequently because the only way to undo a selection is to quit without saving.

 Pan This tool slides the model around the screen. Alternatively, press and hold the mouse's scroll wheel down.

 Dolly With this tool, you, the viewer, move through space to get closer to, or farther from, the model.

Dolly isn't the same as zoom. Dolly moves you, the viewer, closer to the model, passing other items on the way. It gives you a close-up view like zooming in does, but you, the viewer, are moving through space. Movie makers use a dolly for special effects. For instance, they can film a subject that is moving toward or away from them while maintaining the same distance from it. Zoom is when you, the viewer, remain at a constant position, but the camera makes the model look bigger or smaller. Zooming in is like a telephoto lens; zooming out is like a wide-angle lens. You can zoom in the Catch desktop app even though there isn't an tool icon for it; just rotate the mouse's scroll wheel up and down.

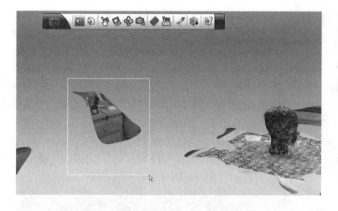
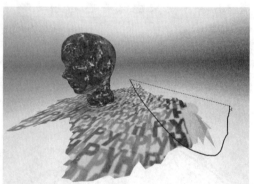
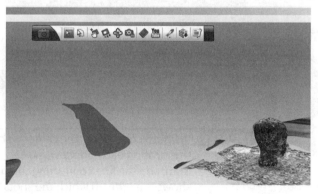

Figure 5-41 Drag the Rectangle and Lasso tools around areas and then release. The selection turns red.

Orbit   This is the 3D navigation mode. You, the viewer, rotate around the model at any height or angle. A circle appears on the work canvas; left-click the mouse inside it and drag to orbit. Left-click the mouse outside it and drag to roll. Rolling is useful to straighten out a skewed model. Another way to orbit is to hold the ALT key and left mouse button while dragging the mouse (press ALT first and the left mouse button second; don't press both at the same time). To get out of Orbit mode, right-click anywhere on the canvas and select Exit Navigation Tool (see Figure 5-42). If you're having trouble orbiting, check whether the Photo lock icon in the toolbar has a blue background (meaning it's active).

 Photo Lock   This is the 2D navigation mode (see Figure 5-43). You, the viewer, move around the capture by jumping from photo to photo (camera view to camera view). This mode is used to study the relationship between the capture and the camera photos. Turn the cameras on at View | Display | Cameras. You can review the locations from which you took the photoset as well as look for outliers and other issues to fix before resubmitting the photoset for a new capture. The green camera represents the first shot taken, and the line connecting all the cameras shows the order in which the subsequent shots were taken.

You can navigate three ways in Photo Lock mode: click each thumbnail, press the left and right keyboard arrows, or click each camera. Hover the mouse over each camera to display the photo it took. That photo will appear behind the model (see Figure 5-44), and the corresponding photo in the photoset bar will

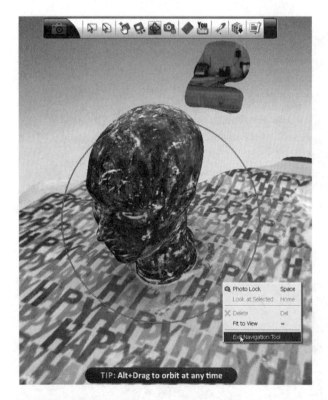

Figure 5-42   Orbit mode. Click inside the circle to orbit, and click outside it to roll. Right-click to exit.

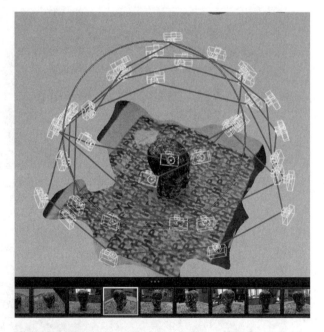

Figure 5-43   In Photo Lock Mode, the cameras show the location each photo was taken from, and the line connecting them shows the order in which the shots were taken. The highlighted picture in the photoset is the current view.

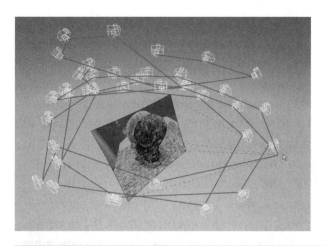 

Figure 5-44 Hover the mouse over each camera to display the photo it took. An outline means you're looking behind the photo.

highlight. If just an outline displays behind the model, you're looking at the back side of the photo (opposite the camera). You can turn the photos off by unchecking Images in the toolbar's Display Settings.

You can temporarily 3D-orbit in Photo Lock mode by clicking the opaque arrows in the lower-right corner to move through the photoset (see Figure 5-45). Outlined arrows mean there's no adjacent photo to move to. Another way is to

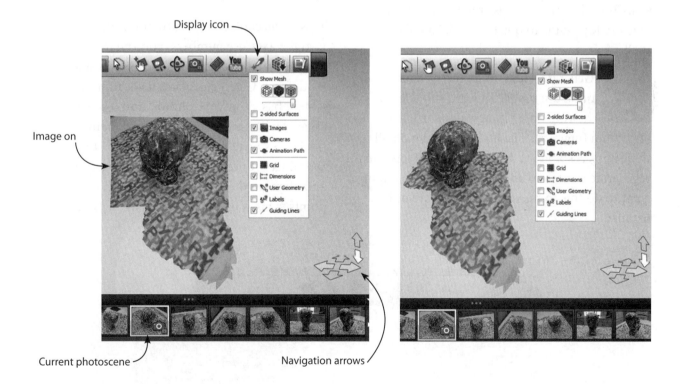

Display icon

Image on

Current photoscene

Navigation arrows

Figure 5-45 In Photo Lock mode you can toggle the camera photos (images) on and off. Orbit by clicking the opaque navigation arrows. The outlined arrows mean there is no adjacent photo to click.

press ALT and the left mouse button (remember to press ALT first). Images that are turned on will orbit along with the model. You can toggle from Photo Lock to 3D Navigation mode in three ways: press the keyboard's spacebar, deselect the Photo Lock icon in the toolbar, or right-click the work canvas and choose Photo Lock.

 Export Video Make a video in .avi format of the whole photoset or of selected keyframes (photos) from it. Catch interpolates between the keyframes to make a smooth video.

 Upload to YouTube Log in to your YouTube account if you have one, and share your .avi file with the world.

 Creation Tools Here you can, among other things, make reference points (see Figure 5-46). These let you scale the capture to a real-world dimension and display dimensions. When working inside another program, such as AutoCAD, you can snap 3D primitives to those reference points. One use for that is to quickly model a building shell.

 Generate Mesh The three densities of mesh are mobile (lowest; suitable for viewing on a phone), standard, and maximum (highest). The first processing on all three apps is standard density. This tool lets you regenerate a mesh to another density. You might want to do this to make the model more or less detailed.

 Display Settings Here, you can toggle the cameras, grid, guidelines, animation path, dimensions, and other features on and off (see Figure 5-47). You can also display the capture three ways: phototextured, wireframe (mesh) and phototextured, or just wireframe. Activate the Photo Lock mode and click the respective icons (see Figure 5-48). The slider underneath them changes the transparency. Viewing phototexture superimposed over mesh is useful to check the camera views because poorly stitched ones become more apparent. The model will continue to display your new choice in orbit mode, but you'll need to return to Photo Lock to change the display again.

If you'd like a grid on the work canvas, toggle it on from the toolbar's Display Settings icon (see Figure 5-49). The squares will appear larger and smaller in tandem with the model.

World Coordinate System

This is the capture's (0, 0, 0) point. Its x, y, and z axes use numbers (coordinates) to uniquely describe a point's location in space. By default, Catch calculates a coordinate system based on your photos, and it's rarely aligned with the capture. You can change the WCS's position by selecting Edit | Define World Coordinate System (see Figure 5-50) and aligning the origin and axes with reference points. This is useful in situations such as

Figure 5-46 Creation Tools

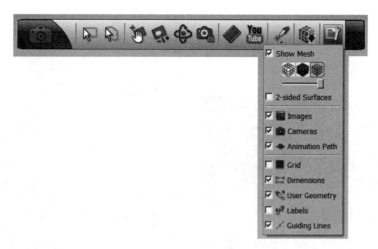

Figure 5-47 Display Settings

Figure 5-48 View the capture as a phototexture, phototexture and wireframe (mesh), or just as a wireframe.

Figure 5-49 Toggle a grid on through the Display Settings tool.

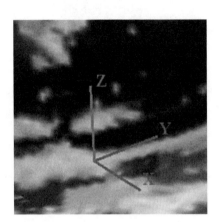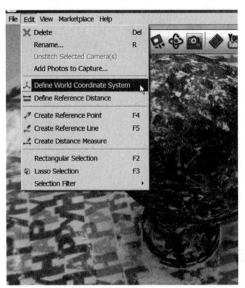

Figure 5-50   The WCS is the capture's origin point and axes. It can be aligned with the capture by reorienting its origin and axes with reference points.

aligning a skewed building capture with the WCS to enable displaying it orthogonally. A 123D Catch channel YouTube video called "Custom Coordinate System" shows how to do just that (www.youtube.com/watch?v=yD1FV-cFzYE&list=PLC04311A173548A12).

Note that there are two WCS icons. One is in the corner and one is in the scene. The corner icon is for reference as you orbit around; the scene icon is for reference within the model (which might not always be visible). You can turn off the one in the scene at File | Preferences | Display | Draw Axis.

Make an Animation

Here are the steps for making an animation (movie):

1. Choose how to display the capture. You can change the background color at File | Preferences | Colors. From the toolbar's Display settings, you can make the capture a wireframe, a wireframe and phototexture, or a phototexture only.

2. Click Create Default Animation Path to choose the whole photoset (see Figure 5-51). The animation path will appear. Clicking the arrow in front of Animation Path toggles it between "show" and "hide".

3. When the animation path is shown, a blue line connecting each camera in the capture appears, showing the order of views. Clicking on the "2s" (two seconds) between each frame brings up a box to change the time between each view (see Figure 5-52). Click the arrow after Animation Path to watch the video. Click it again to stop.

4. Render the animation—that is, turn it into a movie (see Figure 5-53). Click the Export Video tool. Enter the settings you want, click Render, and all those keyframes will be turned into an .avi file. Find the file in the same folder as the .3dp file. It will have the same name plus "movie."

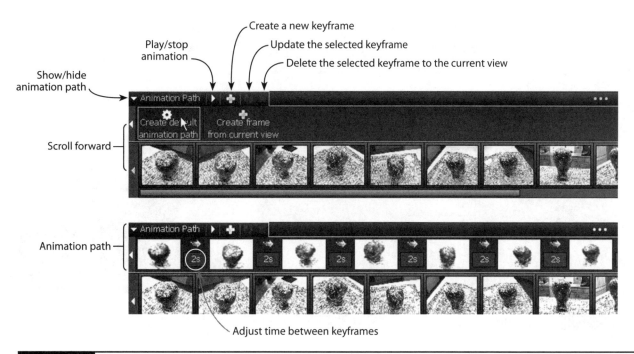

Show/hide
animation path

Play/stop
animation

Create a new keyframe

Update the selected keyframe

Delete the selected keyframe to the current view

Create default
animation path

Create frame
from current view

Scroll forward

Animation path

Adjust time between keyframes

Figure 5-51 Create the default animation path (all keyframes are automatically selected).

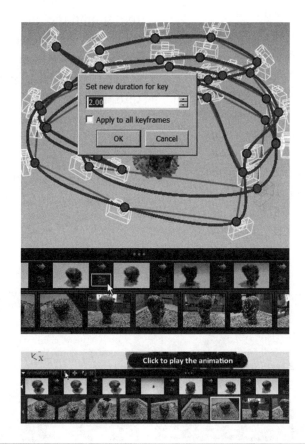

Figure 5-52 The line connecting the cameras shows the order of views. The time between each view can be changed.

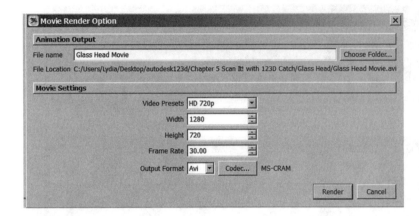

Figure 5-53 Render the animation.

Customize an Animation

We made a default animation, which uses the whole photoset. Maybe you want to remove some photos or add new ones. Add a keyframe by highlighting it and clicking the plus sign. Delete a keyframe by highlighting it and clicking the X. Import a keyframe via the context menu that appears when right-clicking (see Figure 5-54), choose Add Photos to Capture, and navigate to those photos. Notice that the blue line in the work canvas changes to reflect the new animation path (see Figure 5-55).

Because the processing software needs overlapping points to do its job, you'll need to resubmit the catch with the new photos. Alternatively, you can manually stitch them for greater control. You may have also found that some of your existing photos didn't get stitched into the photoscene because Catch couldn't find common features between them and other photos (perhaps they were outliers, or too dark, or just didn't have enough matching points with other photos). Unstitched photos appear darker and have a warning triangle (see Figure 5-56).

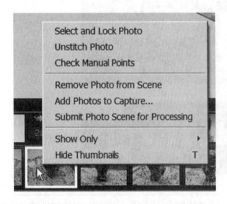

Figure 5-54 Right-click on keyframes to bring up context menus.

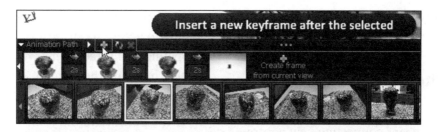

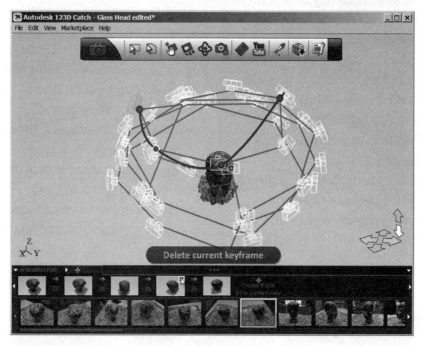

Figure 5-55 The animation path changes with the keyframes.

Figure 5-56 Photos that didn't get stitched into the photoscene are dark and show a warning triangle.

You can manually stitch those, too, if you want them included.

Manually Stitch Photos into the Photoscene

Manual stitching means connecting new photos to ones already stitched into the photoscene (see Figure 5-57). Double-click the photo you want to stitch; it, plus two adjacent views, will appear. Click the wand and then click a specific feature in the unstitched photo. Ideally, this feature will contrast with the background and be easily identifiable, such as a corner. When you click the feature, a yellow circle appears, along with a magnifier to enable more accurate placement. After you click one or two points, an automatic matching process tries to find the same feature in at least one other stitched photo;

when it does, a yellow square appears. Click that square to accept its suggestion or relocate it, and (hopefully) all three points will turn into green circles, meaning they match. Then do this three more times.

If the point remains yellow in all three windows, there isn't enough *parallax* yet (the effect of an object appearing differently when viewed from different angles). Hence, you'll need to select that point in another view; bring one up by clicking the white arrow on either side of the photos. When the point is adequately defined from different photograph viewpoints, the yellow circles will turn green. If the software detects that the points don't match, you'll get a message saying so, and the point will appear as a red square (see Figure 5-58). Right-click the red square to delete it.

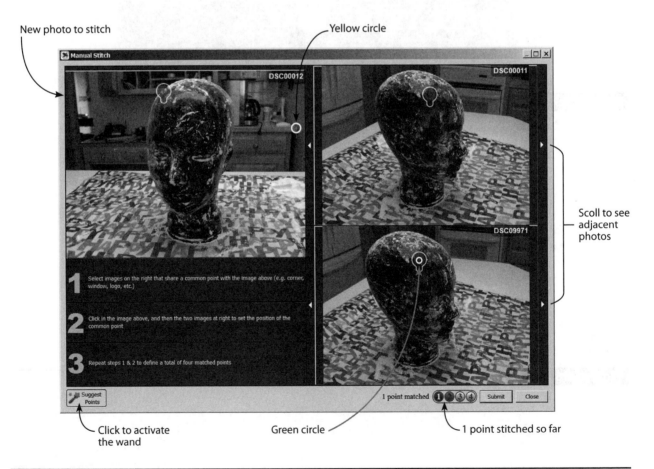

New photo to stitch · Yellow circle · Scoll to see adjacent photos · Click to activate the wand · Green circle · 1 point stitched so far

Figure 5-57   Manually stitch a photo by finding the same point on it in different views.

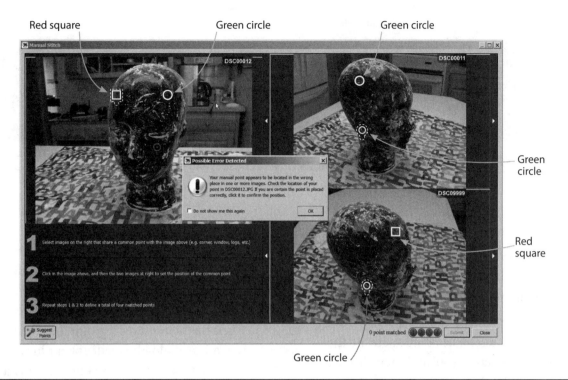

Red square   Green circle   Green circle

Green circle

Red square

Green circle

Figure 5-58 Points that don't match are flagged with red squares. Points that match in all views appear as green circles.

Ultimately, you need four green points across three photos (see Figure 5-59). Finally, click Submit to reprocess the photoscene.

TIP If too many images have warning triangles, or if you want to add too many new photos, it may be easier just to reshoot the whole photoscene. Alternatively, just reshoot the ones that didn't get stitched, using the same locations (turn on the cameras to see those locations).

Let's now use some of the tools discussed earlier. We'll use the Creation tools to make some reference points on the glass head capture, scale it, and add a dimension line. Then we'll use the Generate Mesh tool to change its density.

Make and Delete Reference Points

Reference points are 3D points made by clicking a 2D image. Activate the Photo Lock mode and display the phototexture only, not the mesh. Click open an image from the photostrip thumbnails, and then click somewhere on the image to set the first reference point (again, a magnifier will appear for more accurate placement). Note that a green pushpin appears on the photostrip thumbnail, and green points appear in every other thumbnail (see Figure 5-60). They're all the same point. If you wanted, you could click any of those thumbnails to finesse the location of the reference point you just made.

Figure 5-59 This graphic in the screen's lower-right corner means the photo has been successfully stitched into the photoscene.

Green point

Green pushpin

Figure 5-60 A pushpin appears in the thumbnail image chosen, and the reference point appears as a green square in all other views.

You can click the second reference point on the same image or choose a different image. I chose a side view so as to make it easier to visually line up the second point with the first. Click the second reference point (see Figure 5-61).

To delete reference points, activate the rectangle selection tool and then click each one individually (see Figure 5-62). Don't drag a window, because that selects mesh. When a reference point is selected, it will turn red and a pushpin will appear. Right-click and select Delete or just press the DELETE key.

Scale the Capture

Now that we've got two reference points, go back to the Creation Tools menu and choose Define Reference Distance (see Figure 5-63). Click the first reference point and then click the second. A dialog box will appear in which you can overtype the size you know the distance to be. Then click OK to scale the whole photoscene. Catch is unitless, so the "8" I entered is intended to represent inches, but may also represent feet, meters, Bullwinkles, or whatever you want. After scaling, any distance measured between

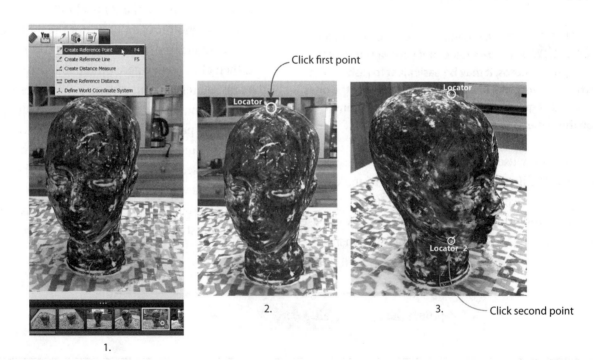

Click first point

Click second point

Figure 5-61 The second reference point was placed on a different photo than the first. This made visually lining them up easier.

1. Select the reference point

2. Right-click to delete

Figure 5-62 When a reference point is selected, it is red and a pushpin appears.

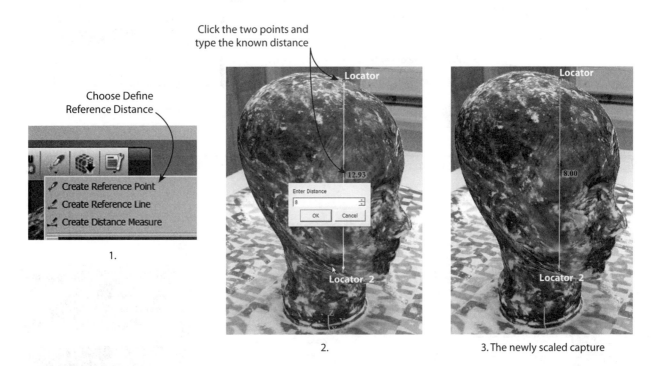

Click the two points and type the known distance

Choose Define Reference Distance

1.

2.

3. The newly scaled capture

Figure 5-63 Use reference points to scale the capture.

reference points will be the same unit you intended the resize to be. If you import this file into AutoCAD, you'll need to define the units.

If you want to scale the image again later, choose Create Reference Distance again. A dialog box will appear, confirming that you want to rescale. Click OK and then repeat the scaling process.

Add a Dimension Line

Click two reference points at the endpoints of a length you want to dimension. From the Creation Tools menu, choose Create Distance Measure.

Then click the reference points. The distance between them will appear on a dimension line (see Figure 5-64).

Generate Mesh

To get a different mesh, resubmit the capture for reprocessing (see Figure 5-65). You'll need to log in to your 123D account (sign in through the Marketplace window). The photoset isn't uploaded again because Autodesk saved it on its servers. As with the first capture, you'll be asked if you want to wait or be e-mailed a link. You'll probably want to choose e-mail when

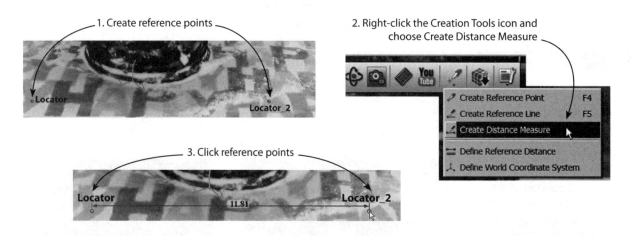

Figure 5-64 Create, and then click, two reference points to place a dimension line between them.

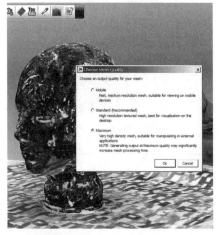
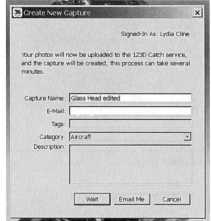
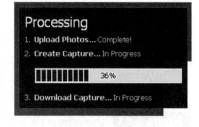

Figure 5-65 Reprocess the capture to change the mesh density.

requesting the maximum density option. If you wait, the new capture will open in the app when it's finished processing. Generate the density you want before making any edits because you'll lose those edits in the reprocess. You can also select just a portion of the mesh for reprocessing. Highlight what you want with the square or lasso tool and start the upload. Catch will reprocess the highlighted part and delete everything else. Figure 5-66 shows the glass head with mobile and maximum densities.

Troubleshoot the Desktop App

You may experience some problems while working in the desktop app. Here are common ones and possible solutions:

Problem:

- The toolbar doesn't appear.

Possible solution:

- Go to View | Navigation and choose a tool. That may bring the toolbar back up.

Problem:

- The photoset gets stuck in transmission.

Possible solutions:

- Check for foreign characters in the directory path.
- Try putting the photoset in a different location.
- Downsize the resolution of the photos.
- Ensure there's no firewall blocking 123D Catch.

Problem:

- You get a connectivity error message even though you know you're connected to the Internet.

Possible solution:

- Create a new project and log in to your account again.
- Reinstall the app.

Problem:

- The desktop app doesn't work.
- The desktop app gives error messages about being in offline mode.
- The desktop app won't load the photoset.
- The desktop app won't let you log in.

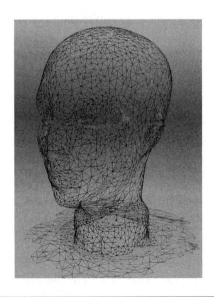 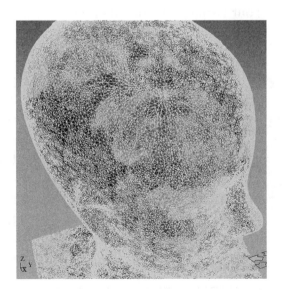

Figure 5-66 Standard and maximum mesh densities

Possible solutions:

- The version you have may have expired. Download the latest from 123dapp.com/catch and reinstall it.

- Try logging in through the web app.

Problem:

- The capture appears small, with too much background.

Possible solutions:

- Examine the cameras for outliers. Click an outlier to highlight the corresponding photostrip image and then delete it. Save, close, and then reopen the file.

Export the Catch

Now let's turn the catch into something besides a .3dp file. You need to be logged into your 123D account to do this. Go to File | Export Capture As. Both free and Premium members have all the choices shown in Figure 5-67. OBJ, DWG and FBX files can be imported into other programs, such as AutoCAD, Rhino, and Maya, for further development (be aware that the DWG format only exports lines and reference points). RZI, LAS, and IPM files are less useful for our purposes because they mostly enable the viewing or exchange of files and 3D point cloud data.

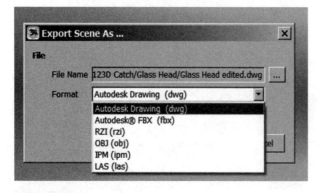

Figure 5-67 You can export a scene into any of the formats shown here.

Export the capture as an .obj, which is a 3D-printable file. Look for it in the same folder as the .3dp file (it will also have the same name). Note that . mtl and .jpg files of its textures were created at the same time (see Figure 5-68). Keep them in the same folder as the .obj file—they will be needed if you plan to import and color-render the .obj in another program, such as Autodesk Maya or 3ds Max.

Just for fun, launch 123D Design and import the .obj into it via 123D's Open menu. Figure 5-69 shows what it will look like. To see its form a bit better, select it and then paint a material onto it.

Tips on Photographing a Subject

The most skillful use of Catch's tools can't compensate for a poor photoset. Great captures are made from great photos, so let's now discuss how to take those photos.

When using consumer-grade cameras and entry-level reality capture software, small subjects are easier to successfully model than large ones because they need fewer photos, less processing time, and less manual repair. However, whether you're using entry-level or professional software, reality capture can only be done on subjects that have a matte, patterned, or textured surface. Reflective, shiny, and transparent subjects—mirrors, glass, and polished metal—don't give the processing software the information it needs. Ambiguous features such as hair, fur, and large swathes of plain fabric, or even symmetrical surfaces, may also pose problems. They often appear in the file as holes or bad mesh. If a large part of the subject is plain, place colored sticky notes with black marker scribbles in the corners of that part. That will help the processor find the common points it needs to correctly

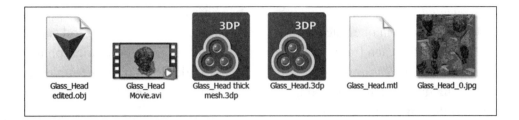

Figure 5-68 Exported files will be placed in the same folder as the .3dp file. Keep all files that are created because they'll be needed to import the model's colors into another program.

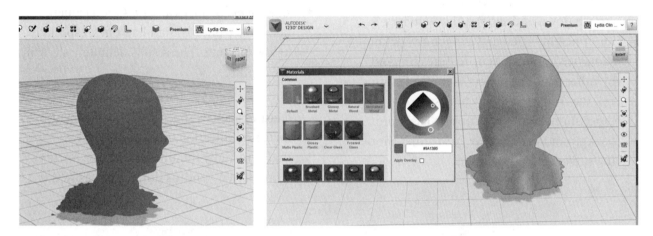

Figure 5-69 The capture imported into 123 Design.

stitch the model together. Glass and other reflective subjects can be captured when covered with flour, talcum powder, or tempera paint (suggested because it washes off easily).

A successful catch requires methodical photographing. Here's how:

- *Place the subject where there is space to move completely around it.* Remove anything blocking your view.

- *Place the subject on an appropriate surface.* Just like the subject, the surface shouldn't be reflective, transparent, or shiny. Choose a surface whose color and pattern contrasts with the subject. Gift wrap, newspaper, magazines, and cotton fabric work well. If Catch mixes your subject and surface up, change the surface or put small items under the subject to raise it a bit above the surface.

- *Illuminate the subject with diffused light.* Consistent, even lighting all around the subject is critical. Best results come from shooting the subject in full shade, or on an overcast day, or indoors. Avoid stark shadows, spotlights, strong backlighting, and direct sunlight on the subject, because all this results in under- or overexposed photos. Also avoid highlights, because they move around the subject as you move around it. Outdoor lighting conditions vary with time of day and year; the middle of a bright summer day probably won't work, but early evening may.

- *All photos must be identically lit.* Don't use a flash, because it exposes (illuminates) each photo differently. Camera phones have an automatic exposure feature, making their flash photos especially bad.

- *Move 360° sequentially around the subject* (see Figure 5-70). Don't stand in one place and just turn the camera. Fill each frame with the whole subject. Take 15 to 20 shots every 20° or so, maintaining the same distance from the subject, and the same distance between shots. Shoot a second sequential circle at a different height—ideally, top-angled down. Again, maintain a consistent distance from the subject and between shots. If needed, crouch down and shoot a third circle at a lower height. Then move in for any detail shots needed, such as of features blocked by other features, or very complex features. Fill the frame with the whole subject on those shots, too. If the capture has more background than subject, you didn't shoot the subject closely enough.

- *Keep the subject still.* All features need to be in the same place for good stitching. Don't move or lift it to photograph hidden areas or underneath it. If the subject moves even a little, start over.

- *Overlap the photos.* Every feature should appear in at least three or four different photos taken at varying angles. Missed spots will appear as holes in the model.

- *Use a good quality camera.* Good cameras take better photos than bad ones, especially in low lighting. The results definitely show in the capture.

- *Take raw images.* Turn off camera features such as sharpness enhancement, image stabilization, and anything else that artificially modifies the photo. Such "interpreted" photos don't stitch as well.

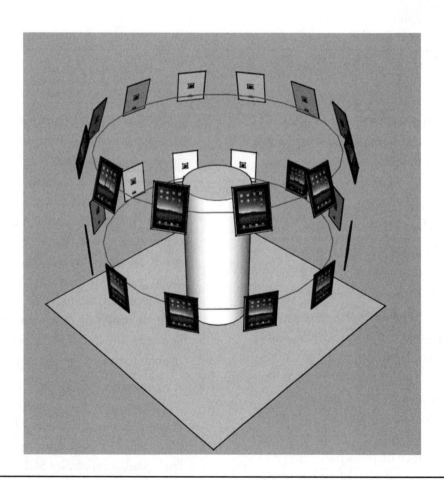

Figure 5-70   Move sequentially around the subject, taking photos at regular intervals.

■ *Focus.* Sharp images are critical because blurry ones don't provide the feature information needed. If you change the aperture setting to blur the background via depth of field, do that in all the photos. An 18 mm focal length is a good setting. Hold the camera against your body for extra stabilization.

■ *Resolution.* Resolution is the number of pixels per inch. A higher resolution provides more detail in the model. However, Catch will resize anything larger than five megapixels.

■ *Orient all photos similarly.* The photos need to be either all portrait or all landscape.

■ *Upload the photos as is.* Don't crop or edit. You can downsize their resolution to speed up processing, but that will affect the model's detail and quality.

■ *Upload the proper number of photos.* Between 70 and 100 photos gives best result; more freezes the software. Don't upload more than 70 in the iPad/iPhone and web apps, or over 100 in the desktop app.

■ *Delete bad photos before uploading.* Blurry, overexposed, or underexposed photos mess up the stitching process.

■ *Get best results when photographing a person* by seating them and telling them to avoid small facial movements by looking forward, blinking between shots, and closing their mouth.

Finally, study models in the 123D Gallery. What subjects make the best models? Do some look light on one side and dark on another? That's the result of uneven lighting. Are there holes or messed-up areas? That's due to features that were unclear, not captured enough, or not captured at all. When surfaces are included, note what was used. After your own model is generated, study the photoset for clues on what to do differently.

Summary

In this chapter, we used 123D Catch's mobile, web, and desktop apps to create and edit a photorealistic model called a *capture*. We made one capture with an iPad and one with a digital camera. Then we made an animation and .obj file of a capture and discussed how to scale it. Captures jump-start the design process because they enable you to start a project with existing data, instead of having to create it all from scratch.

Sites to Check Out

■ **The 123D Catch YouTube channel** www.youtube.com/user/123DCatch. Includes lots of tutorials.

■ **Smithsonian X 3D Project** http://3d.si.edu/

■ **Zebra Imaging's holographic images** www.youtube.com/watch?v=Xp7BP00LuA4. These images are created from 3D models.

■ **NASA's downloadable 3D models** www.nasa.gov/multimedia/3d_resources/assets/helmet.html.

■ **Capture and 3D-print a bobble head doll of yourself** www.instructables.com/id/3D-Printing-your-own-full-color-bobblehead-using-1/.

■ **Short movie made via reality capture** www.youtube.com/watch?v=TSOJl8SNZLM.

■ **Meshlab** www.meshlab.net. A website that processes and edits reality capture photosets for free.

■ **Photosynth** www.photosynth.net. A free reality capture program by Microsoft.

CHAPTER 6

Mix It Up! with 123D Meshmixer

.STL AND OBJ FILES ARE mesh models inside an invisible shell. The Tinkercad and Design apps let you make superficial edits to the shell such as scale and move it. Meshmixer, on the other hand, puts you inside the shell to edit the mesh itself. Specifically, you can perform the following tasks:

- **Edit a mesh model**   You can change a mesh model's density, alter it with brushes and other tools, combine multiple models, and even heal open mesh models into closed, watertight models.

- **Create simple 3D models**   You can create printable models from scratch.

- **Prepare a model for 3D printing**   You can thicken a model's walls, generate supports, orientate it on the build platform, and then send it directly to your printer or a third-party service bureau. All the 123D apps send their files to Meshmixer for printing preparation. Meshmixer has replaced the Autodesk Print utility (the software formerly used for this purpose).

This robust app has two parts: Modify, where models are healed and altered, and Print, where models are prepped before sending to a personal 3D printer or service bureau. This chapter shows how to use tools in the Modify part to make simple alterations. Prepping files for 3D printing is covered in Chapter 9.

 To get started, download Meshmixer at www.123dapp.com/meshmixer and then install and launch it. If you want to play along with the tutorials, you can find the donut, glass head, and Pac-Man and Pinky files in the 123D Gallery by performing a search on Lydia Cline.

The Meshmixer Interface

Upon opening Meshmixer, you'll see the launch screen, which shows the workspace, six large square buttons, a horizontal menu bar at the top, and a vertical menu panel on the left side (see Figure 6-1).

The six square buttons offer the following functionality:

- **Import**   This button brings in four file types: .stl, .obj .ply, and .amf. You'll find that the .obj files work the best because they're the most robust and best defined of all the formats (not as loose or open ended). Plus, they contain any color information.

- **Open**   This button brings in .mix (meshmixer) files.

- **Bunny, Sphere, and Plane**   These buttons import premade files for you to experiment with.

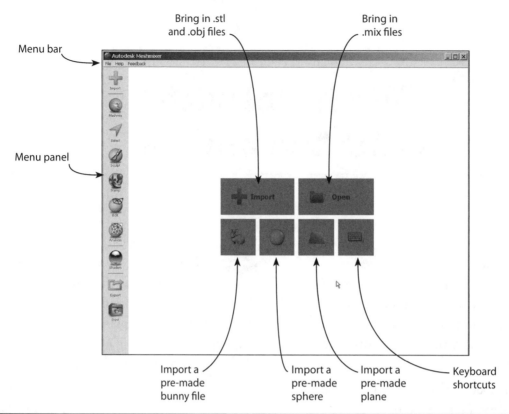

Figure 6-1 The launch screen

- **Keyboard Shortcuts** This button brings up charts of hotkeys (shortcuts) for navigating the workspace, selecting, and activating tools. The navigation keys are different from the other 123D apps.

The menu bar contains links to help sites and a feedback field. It has different options after a file is opened. Right now, clicking the File menu accesses the same files as the square buttons. However, note the File | Import Textured option, shown in Figure 6-2, which you can use to bring in a model's colors.

The menu panel contains icons that access the navigation browser, parts libraries, selection tools, sculpting brushes, manipulators, analyzers, display modes, and the 3D print area. Some of these tools work similarly to those in the Design desktop app and the Catch web app.

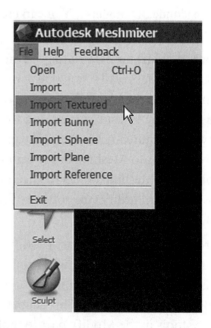

Figure 6-2 The File | Import Textured option brings in an .obj and its associated color files.

Import a Textured Donut Model

This project uses a capture (model) of a donut I made in 123D Catch When you submit a capture for processing, .obj and .stl files are automatically created. Download these files by logging in to your 123D account, clicking the model's thumbnail, and selecting Edit | Download | Download 3D Models (see Figure 6-3).

Select File | Import Textured, navigate to the .obj file (see Figure 6-4), and double-click to import it. Meshmixer will read the .mtl and .jpg files that are downloaded with the .obj file

1. Click on thumbnail

2. Click here

3. Scroll to, and click here

Figure 6-3 Steps for downloading .stl and .obj files from your 123D account.

mesh.obj   tex.mtl   tex_0.jpg

Figure 6-4 Only the .obj file is imported into Meshmixer. But for it to include colors and textures, the .mtl and .jpg files must be in the same folder as the .obj at the time of import.

upon import. Make sure they're all in the same folder. If the texture doesn't appear after the file is imported, you may need to drop the texture shader onto it. We'll talk about that later.

Figure 6-5 shows the imported donut model. The workspace now displays a grid. When orbiting around the model you'll see a small, red sphere. It's the grid's origin (0,0,0),

and is the axis of rotation when you orbit. So it's best to model on or near it. The File menu has a Help submenu that brings up a short tutorial on the activated tool (see Figure 6-6). Also, the Object Browser box, which lists the models in the file, appears in the lower-right corner (see Figure 6-7).

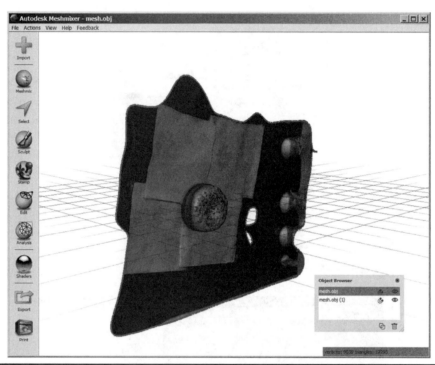

Figure 6-5 The imported donut model

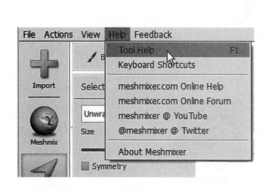

Figure 6-6 Access a tutorial on an activated tool via File | Help.

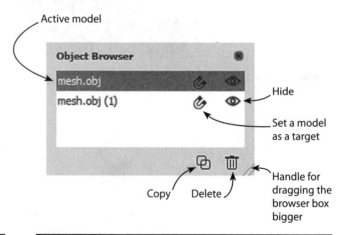

Figure 6-7 The Object Browser box lists the models in the file.

The Object Browser Box

The Object Browser box lists all the models in the workspace. The highlighted entry is the active model, meaning it can be manipulated and edited. It displays brighter than the nonhighlighted ones. Two models are shown even though we just imported one. The File | Import Texture feature is experimental at this time of writing and can be glitchy. Importing this model as two is a glitch.

The Object Browser box has some graphics. The eye displays and hides a model, the magnet pulls a selected model toward a target model, the two squares copy a model, and the trash can deletes a model. If some graphics aren't visible, drag the box bigger from its lower-right corner. If the Object Browser box doesn't appear at all (or disappears), bring it up from the menu bar via View | Show Objects Browser.

Navigate the Workspace

Figure 6-8 (left) shows the hotkeys used to tumble, pan, and zoom. Tumble rotates around the point that you (the camera) are looking at. Pan slides the model around the screen, and

Zoom makes your view closer to or farther from the model. Hold down the spacebar to access the Hotbox, a transparent window that's shown on the right in Figure 6-8. It has icons that let you tumble, pan, zoom, and focus (focus fills the screen with everything in the workspace). There's also a square for setting the workspace's background color. We'll discuss other Hotbox graphics later. While holding down the spacebar, press and hold the mouse's left button on a Hotbox icon to activate its function.

You can also tumble with the mouse's scroll wheel, or by clicking View | Config | Free Rotate in the menu bar. If you prefer 123D Design's navigation system, go to the menu bar and click View | Navigation Mode | 123D.

Rotate the Donut with the Transform Tool

The donut's position is vertical because Catch and Meshmixer don't have the same world coordinate system (WCS). To rotate the donut parallel to the grid, first click it to select it. You can also select a model by clicking it in the Object Browser box. Next, click

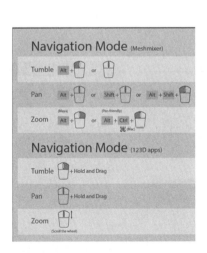

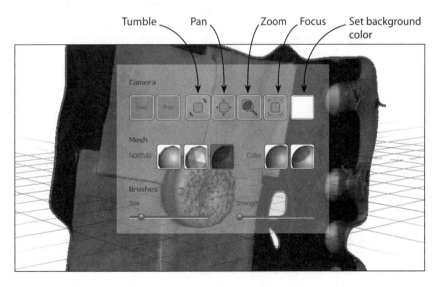

Figure 6-8   Navigate with hotkeys (left) or with Hotbox icons (right).

the Edit icon in the menu panel. This brings up a submenu; click Transform. You can also access the Transform tool directly by pressing T. A manipulator appears (see the first two graphics in Figure 6-9).

The manipulator consists of colored arrows, planes, squares, arcs, and one white square. With it, you move, scale, and rotate the model about its three axes. You'll probably need to tumble the model to see all of them. They're red, green, and blue to coordinate with their respective axes. Click an arrow to move the model up and down along that axis. Click a plane to move it within that plane. Click a square to scale it non-uniformly along one axis. Click the white square at the arrows' intersection to scale the model uniformly along all three axes. Click an arc to rotate the model along that axis (see Figure 6-10). Click the ESC key to cancel an operation. Click Accept in the upper-left screen to keep a transformation. Select Actions | Undo from the menu bar to undo an operation one at a time.

The widget also has a circle with four lettered buttons. *S* is snap; click it to bring up

Figure 6-9 The Transform tool brings up a manipulator widget.

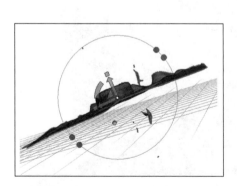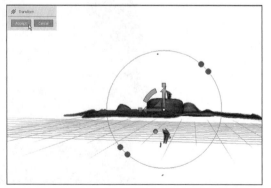

Figure 6-10 Rotating the donut with the Transform tool. When finished, click Accept. Click Actions | Undo/Back to undo an operation.

a numbered axis along which the model will snap. *L* snaps the model along the local WCS (which is based on the imported model, and can be changed with the Transform tool). *W* snaps the model along the global (grid-aligned) WCS. *A* snaps along absolute coordinates.. Snapping only works for moving and rotating, not scaling.

To increase the snap numbers, hold down the left mouse key on an arrow, square, or arc and press the keyboard's UP ARROW or RIGHT ARROW key. To decrease the snap numbers, press the keyboard's DOWN ARROW or LEFT ARROW key. Pressing the arrow keys multiple times increases/decreases the snap numbers incrementally (see Figure 6-11).

Analyze the Donut Model with the Inspector Tool

Although this model is an .obj file, which is a format for 3D printing, it is not printer ready. First, it must be "watertight," meaning if its interior were filled with water, none would leak out. Looking under the donut, we see it has no bottom, and the surrounding surface has large, visible holes, as shown in Figure 6-12. Any hole, large or small, makes a file unprintable. Meshmixer outlines all holes in blue.

In the menu panel, click Analysis | Inspector. Colored pins attached to each hole may appear (the top graphic in see Figure 6-13). A blue pin means there's a hole in the mesh. A red pin means the area is *non-manifold* (more than two planes are attached to one edge). A magenta pin indicates small, disconnected parts. Choose a fill type (minimal, flat, or smooth) from the drop-down box. You can click Auto Repair All and Done, or left-click on a pin to heal its hole individually. This may yield a better result than Auto Repair All, as it lets you customize the fill. If the pin turns gray after you click it, the repair failed. Alternatively, you can right-click a pin, which selects the defect

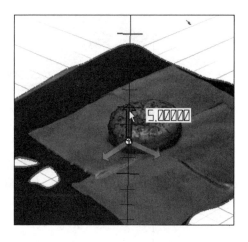

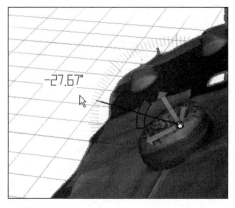

Figure 6-11 Press the arrow keys to increase or decrease the snap numbers.

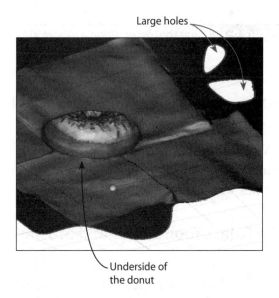

Large holes

Underside of the donut

Figure 6-12 The underside of the donut. It has no bottom, and the surrounding surface has large holes.

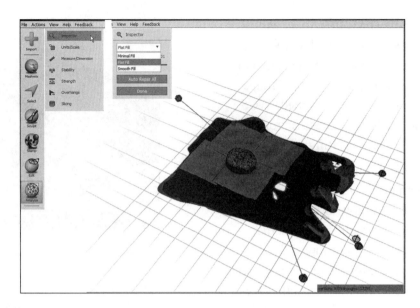

Figure 6-13 Run the Inspector on a model to find and fill holes. The bottom graphic shows the large holes filled after Auto Repair All was selected.

and exits the Inspector so you can perform other operations on that area.

We're not fixing these holes; rather, we're going to remove everything around the donut. For this we need the Select tool.

The Select Tool: Lassos and Brushes

Mesh must be selected before editing. Meshmixer selects faces (individual polygons). You cannot select precise patches by entering numbers; rather, you select with a brush or a lasso. From the menu panel,

click the Select icon or press s. This brings up an options box (see Figure 6-14). The brush is the default tool; to use the lasso, click the button next to the lasso icon. The Size slider adjusts the tool's size, and Crease Angle makes it stop at sharp angles or creases, which is handy when you're trying to brush along the bottom of the model without running up the sides.

Lasso

Click the lasso on either the model or the workspace. Hold down the left mouse key and

Brush
selected

Lasso
selected

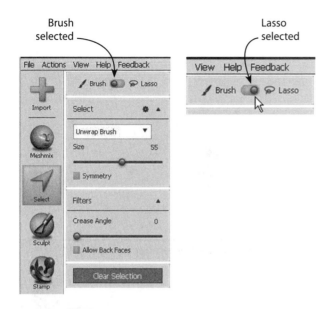

Figure 6-14 The Select tool has brush and lasso options, with the brush as default. Choose the lasso by clicking it.

drag a freehand loop around an area (see Figure 6-15). You must close the loop; that is, you need to end the lasso at its start point. Otherwise, nothing will select. Selected mesh turns brown. If you don't like the selection, press the ESC key or click the Clear Selection button at the bottom of the options box. Hold down the SHIFT key to deselect portions. Press the DELETE key to erase the mesh. Figure 6-16 shows a lasso loop around the bits that imported as a second

model. Remember to highlight that model in the Object Browser box before trying to select it. After the first lasso is made, the Select tool returns to the brush.

When you start and stop the lasso entirely on the model, the lasso appears inside the selection as a boundary loop (see Figure 6-17). Further operations will only affect the mesh inside that loop. If you start the lasso on the workspace, cut across the model, and stop it on the workspace at the other end, you've drawn a *laser line,* and everything on one side will select (see Figure 6-18). You don't have to close the loop. If you place the second point somewhere on the model, you'll still get a selection, but with less predictable results. Only one laser can be drawn at a time; a second laser will replace the first.

If you draw lines instead of drag them (that is, you click the lasso on the workspace, release the mouse, and click it in a second spot), the line in between the clicks will be perfectly straight. You can multiclick a lasso this way. You'll still need to make a closed loop to select anything, though (see Figure 6-19).

When a selection has been made, submenus appear, each containing sub-submenus. Choosing Select | Modify | Invert reverses a selection (see Figure 6-20). Roll your mouse

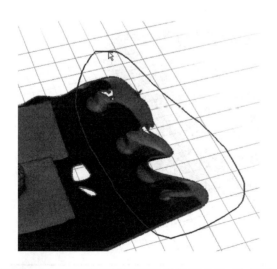
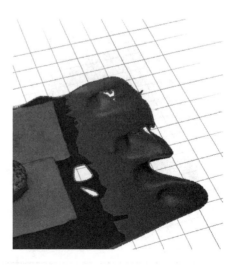

Figure 6-15 A lasso selection must be a closed loop.

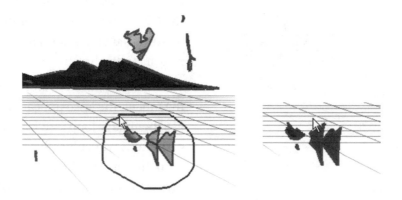

Figure 6-16 A lasso loop around bits of mesh that imported as a second model. Remember to highlight that model in the browser box before selecting it.

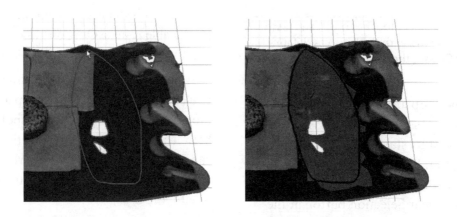

Figure 6-17 A lasso loop dragged on the mesh creates a boundary line inside the mesh.

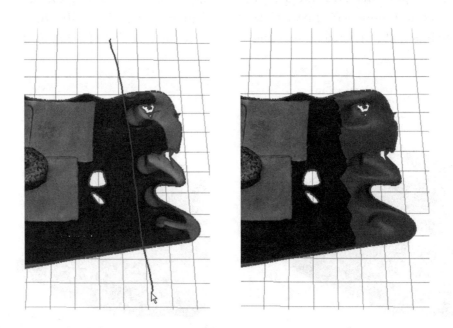

Figure 6-18 A laser line selects one side of the model.

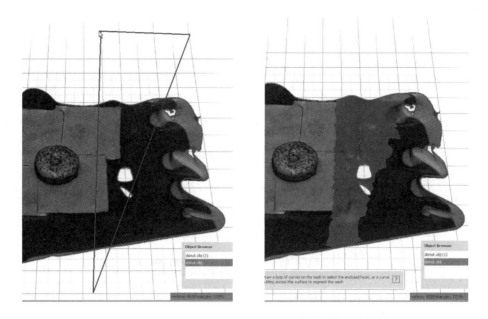

Figure 6-19 A lasso that is drawn instead of dragged makes perfectly straight lines.

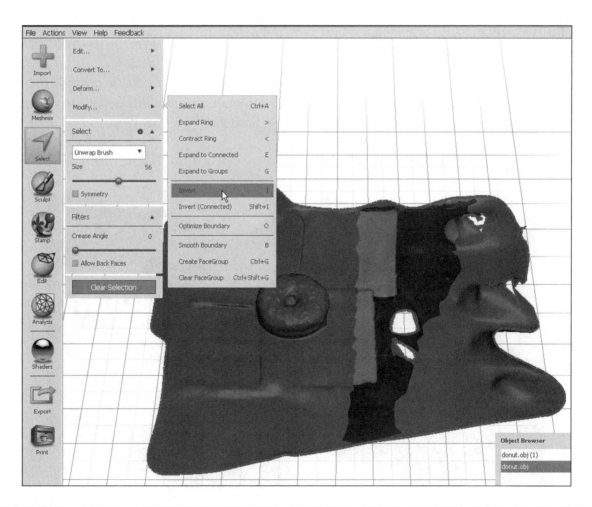

Figure 6-20 The Select submenus have multiple sub-submenus. Shown is Select | Modify | Invert, which reverses a selection.

over the submenus to see all their options. You can expand, contract, and offset a selection. You can brush a small dab onto the model and click Select All, which does just that: selects everything in the highlighted model, whether the pieces are connected or not. To just select one piece, brush a small dab on it and choose Expand to Connected. Figure 6-21 shows the hotkeys for some of those options.

Brush

The brush selector looks like a dark paint spot. Click it onto the mesh to paint with it (see Figure 6-22). There are three ways to make the spot larger or smaller: you can roll the mouse's scroll wheel, move the slider in the Hotbox, or move the slider in the Select options menu. Hold the SHIFT key down to deselect. The Sphere Brush option selects in all directions (see Figure 6-23). Press the DELETE key to erase the selected mesh.

Like the lasso, the brush selector has more options in the sub-submenus. In Figure 6-24, I brushed a selection onto the surface and then

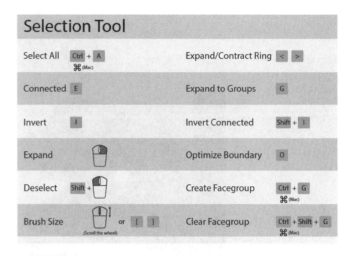

Figure 6-21 Shortcuts for popular selections.

clicked Select | Modify | Smooth Boundary and Accept. The third graphic shows the ragged boundaries turned smooth. Smooth Boundary is a frequently used function, so you should understand its options:

- **Smoothness** This option controls the amount of smoothing applied to the selected border.

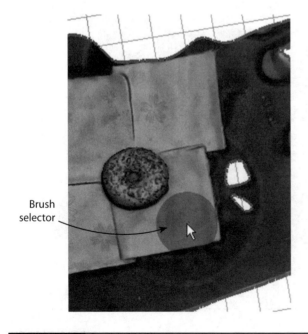
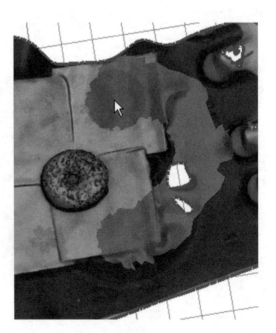

Brush
selector

Figure 6-22 The brush selector looks like a dark paint spot.

Figure 6-23 The sphere brush selects in all directions

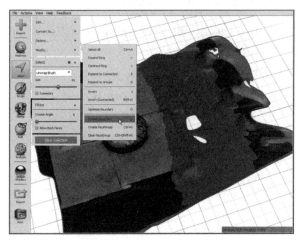

1. Choose the Smooth Boundary option

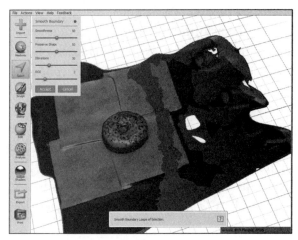

2. Move the sliders

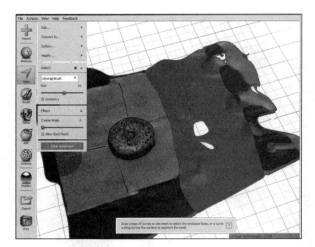

3. The smoothed boundaries

Figure 6-24 Smoothing the boundary lines of a selection

- **Preserve Shape** This is an inflation factor that counteracts the smoothing in order to preserve the selection's overall shape.

- **Iterations** This option controls the rounds of smoothing: the more iterations, the more smoothing.

- **ROI** ROI stands for Region of Interest. It's the size of the support area around the selection's boundary. Mesh is relaxed in this area.

The other selection options are Expand Ring and Contract Ring, which move the selection boundaries incrementally, and Select All, which selects the whole model, including disconnected pieces. Experiment with all these options! In Figure 6-25, I choose Select | Deform | Transform and bent the selection with the manipulator widget. Note the Harden option in the second graphic. It adjusts the sharpness of the transition at the edges.

Facegroups

You might have noticed the reference to facegroups in the hotkey list shown earlier in Figure 6-21. A *facegroup* is an arbitrary selection

of polygons that can be operated on as a whole. Making one lets you return to a specific selection, which is especially useful when paired with the Select | Modify | Optimize Boundaries and Select | Modify | Smooth Boundaries options (you can maintain edges between groups). Facegroups are generated via Select | Modify | Create Facegroup. When the Select tool is activated, simply double-click a facegroup to select it. You can make it a different color by going to the menu bar and clicking View | Mesh Color Mode | Group Color, which enables you to draw facegroups. Click Vertex Color if you're painting the model.

Heal the Donut Model

Let's heal this donut now, pulling together some of the options you've learned so far:

1. Select the donut (see Figure 6-26) and then click Modify | Smooth Boundary. Experiment with the options until the perimeter is as smooth as you can get it and then click Accept. Then click Modify | Invert and press the DELETE key. You're left with just the donut.

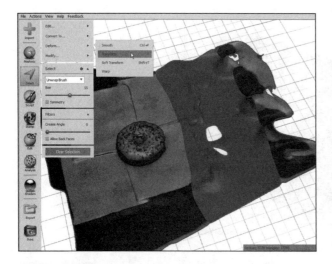 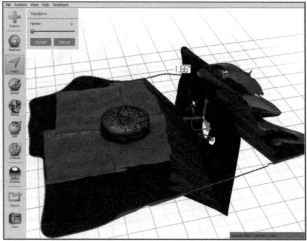

Figure 6-25 Deforming a selection boundary

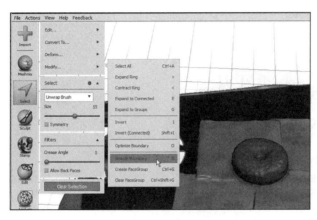

1. Select the donut.

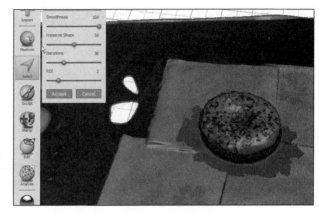

2. Smooth its boundaries.

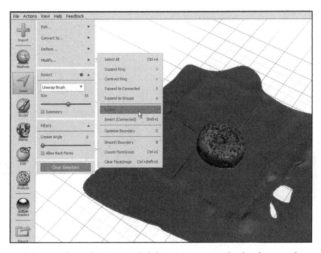

3. Invert the selection and delete to remove the background.

Figure 6-26 Removing background with an inverted selection.

2. Click Analysis | Inspector. Right-click the pin pointing to the donut's underside. This will select its perimeter and exit the Inspector.

3. Click Edit | Erase and Fill (see Figure 6-27). This erases the selection and fills the space with new mesh. Its advantage is an options box, which you can use to finesse the result, such as reduce or increase a bulge. It also removes problem mesh and replaces it with good mesh.

Sometimes Select and Fill doesn't work, and just returns a jagged red area, as shown in Figure 6-28. That area indicates *non-manifold* edges, which are edges bordered by more than

two polygons. This may have occurred from an overcorrection of the perimeter; at any rate, Select and Fill can't fix it. You can undo all the way back to the pre-edit condition and then edit it again. Alternatively, you can select Analysis | Inspector, click the area, and choose Auto Repair All (see Figure 6-29). Although this always works, its disadvantage is you can't refine the result.

Scale the Donut Model

What size should the donut be? How about appropriate for an American Girl doll? 3D printing food accessories for popular dolls

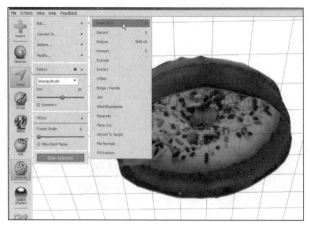

1. Choose Erase and Fill.

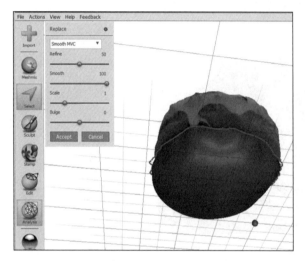

2. Refine the resultant bulge.

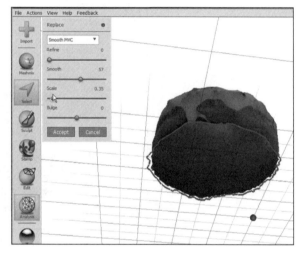

3. The adjusted bulge

Figure 6-27 Edit | Select and Fill will fill the selection (the perimeter of the underside, which was selected via the Analysis | Inspector tool). You can refine the result with the options box.

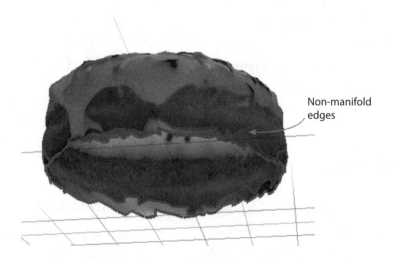

Non-manifold edges

Figure 6-28 The Select and Fill tool can't fix this capture's non-manifold edges.

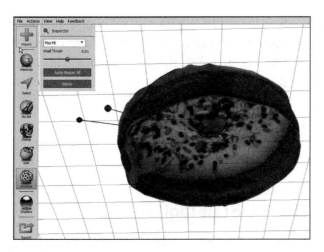
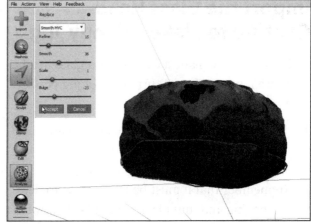

Figure 6-29 The Inspector tool's Auto Repair All option fills the bottom, but has no refining options.

could be a fun project. Therefore, we'll make the donut 2" wide (see Figure 6-30). To start, click Analysis | Units | Scale. A box appears around the model. Snap its arrows back and forth (remember, you can increase the snap precision by using the arrow keys) or type the new dimensions in the X, Y, and Z text fields.

Next, set the units. Type a number in one text field and the numbers in the other fields will scale proportionately. When you change units, a dialog box appears asking if you want to keep the dimension numbers the same (for example, turn 10" into 10 mm) or convert the dimension number to its equivalent in the new unit.

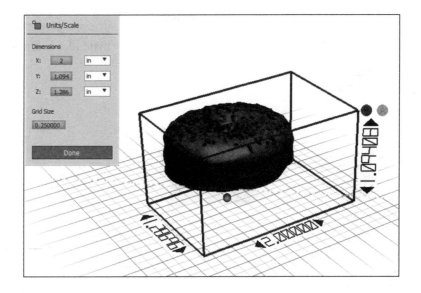

Figure 6-30 Scale the donut via Analysis | Units | Scale.

Import the Glass Head Capture to Make a Candy Jug

In Chapter 5, we made a capture of a glass head. Here, we'll import it and turn it into a jug perfect for holding Halloween treats:

1. Import, select, and rotate the glass head model. Using File | Import Textured, navigate to the glass head's .obj file (remember that it must be in the same folder as the .jpg and .mtl files). Figure 6-31 shows that it imported sideways. Because selecting all the pieces with the lasso is difficult, I selected one of the floating pieces with the lasso and then clicked Modify | Select All. All the pieces then selected.

2. Press T and drag the arc until the model is parallel with the grid. Click Accept. If you can't see the grid, the model probably imported way above or below it; move the model up or down with the blue Transform arrow, and eventually the grid will come into view. Alternatively, rescale the model much larger, using the Transform tool's white square.

3. If, once you've rotated your model, you decide you want to rotate it again, a quick way to do this is with Edit | Align. Note the options for aligning with the different axes (see Figure 6-32). If the workspace grid distracts you, go up to the menu bar and uncheck View | Toggle Grid.

4. To erase the floating parts, draw a lasso around each part, and when it turns brown, press DELETE.

The Plane Cut Tool

To find the Plane Cut tool, click Edit | Plane Cut. This tool divides the model at any location and angle you want. It can cut through the whole model or just through a selection. The Plane Cut tool has Fill and No Fill options, meaning you can make the result solid or hollow. We'll slice the bottom of the glass head off, which will simultaneously level it and remove all extraneous surface. This is easier and more efficient then trying to select and erase all that surface.

When you click Edit | Plane Cut, you can see that the cutting plane has entered the workspace vertically (see Figure 6-33). Rotate it 90° by dragging the widget arc. Then drag the vertical arrow down to move the cutting place down. It doesn't matter that the vertical arrow points up; instead, look at the thick, short arrow to the right.

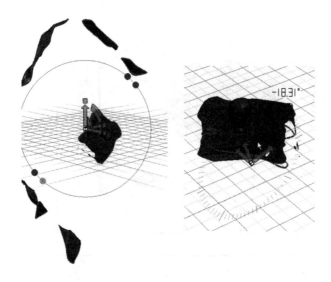

Figure 6-31 Import, select, and rotate the model with the Transform tool.

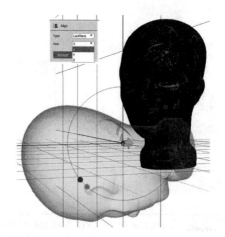

Figure 6-32 Using Edit | Align, you can quickly rotate the model to align with the x, y, and z axes.

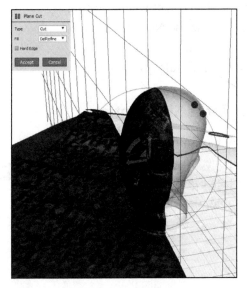

1. Activate Plane Cut.

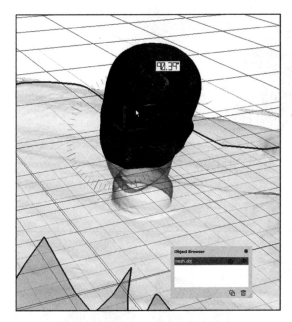

2. Rotate the plane 90°.

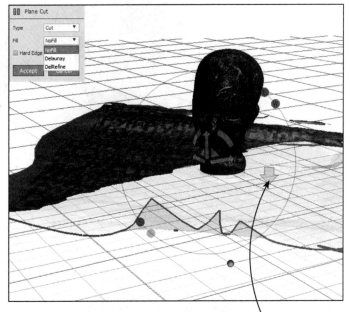

3. Move the plane down.

Short, thick arrow points to what will get cut off

Figure 6-33 Cut the bottom of the head and its surrounding surface off with Edit | Plane Cut.

Everything in front of its point will get cut off, so that's the one you need to orient when rotating the cutting plane. Choose the No Fill option to make the head hollow, thus enabling it to hold candy (see Figure 6-34). Finally, click Accept.

The Plane Cut Tool's Slice Option

The Plane Cut tool has an interesting option called Slice. I've opened a sphere in a new file to demonstrate it (see Figure 6-35). When Edit | Plane Cut is activated, the cutting plane appears

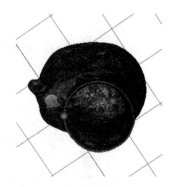

Figure 6-34   Choose the No Fill option to make the head hollow.

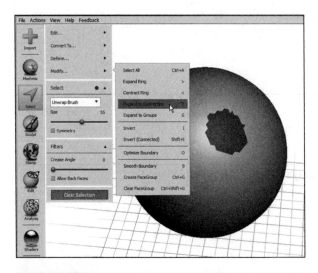

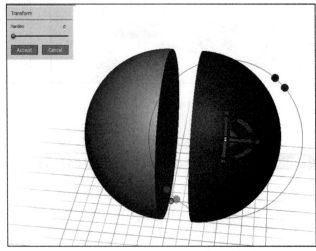

Seam

Figure 6-35   The Plane Cut tool's Slice option preserves both sides.

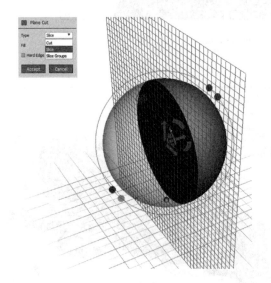

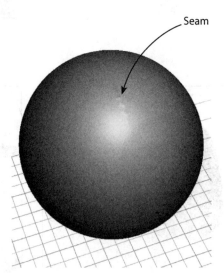

for positioning. Clicking the Slice option and then clicking Done results in a seam at the slice location. Here, I've selected one-half the sphere by select-brushing a spot on it and then clicking Modify | Expand to Connected. I then used the Transform tool to move the slice.

Select and Smooth

There's more to the Select tool than what we've covered so far. For instance, you can select the whole model at once and deform it. Here are the steps to follow:

1. Smooth the head by clicking Select | Edit and moving the Size slider all the way to the right (see Figure 6-36a). This generally creates a border around the whole model. Click the model to select it. If the border doesn't appear after you move the slider, click once on the model, which should simultaneously make the border appear and select it. You can make the brush smaller by moving the slider to the left, but that isn't needed here.

2. While still in Select mode, go to Deform | Smooth (see Figure 6-36b). The default result is good enough for our purposes, but you can adjust it with the options box. Press the ESC key to accept the smoothing and exit.

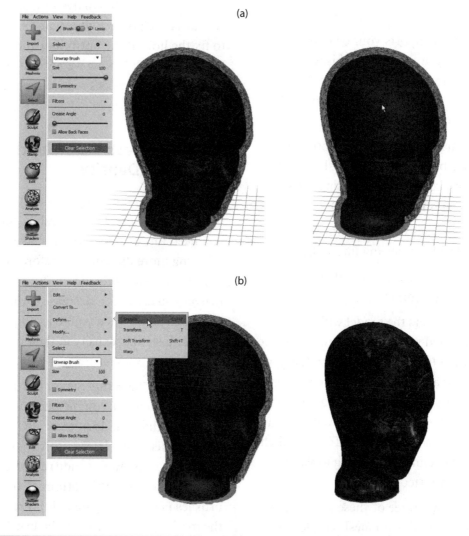

Figure 6-36 (a) Select the whole model by moving the Size slider to the right and then clicking the model. (b) Smooth the head with the Deform | Smooth option in the Select menu.

Change Face Texture with the Shader Tool

The face's black-marbled texture makes it hard to clearly see. Some face textures can make the results from specific operations invisible. If you can't see something on the model, it may be due to its current face texture. You can change the face texture to one of Meshmixer's built-in ones, or you can import your own. In the menu panel, click Shaders and scroll through the options (see Figure 6-37). Click and drag one onto the model and then release. The head will be covered with the new texture. You can also select some face styles through the Hotbox.

Here are the shaders of particular interest:

- **Plus sign** This imports your own texture. It's a bit wonky at the time of this writing; the imported texture wraps around the model, which itself becomes a mirror and reflects the texture.

- **Default** This is a generic gray texture. It makes a nice background for mesh display, which you make visible by pressing w (to turn off the mesh, press w again).

- **Clear** This makes the model transparent, which is helpful for viewing internally.

- **Red tip** This highlights areas that need support during printing.

- **Globe** This covers the model with the texture imported with it. If you imported the .obj file with a texture (conveyed by the .jpg and .mtl files) and the model doesn't display it upon import, drag the globe slider onto it to make the texture appear. Be aware that as of this writing, a File | Import Texture glitch causes some healing and editing tools to remove the imported color.

Figure 6-38 shows three of these shaders: the default (and the default with mesh visible), clear, and red tip.

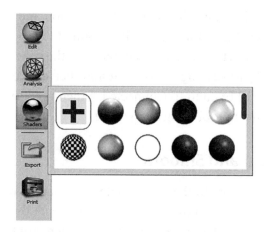

Figure 6-37 Drag one of the Shader tool's options onto the model to change its face texture.

Finally, return to Analysis | Inspector | Auto Repair All. Choose Flat Fill and click Done to fix the hole at the bottom of the glass head (see Figure 6-39). Now we need to cut a hole in the top—but first, let's reduce the mesh density (polygon count).

Make the Mesh Visible and Reduce Its Density

There are too many polygons in this model for our purposes here. Dense mesh is good when you need lots of detail, but it makes careful selecting more difficult. Therefore, if you don't need the detail it provides, reduce it. Another reason you'd want to decrease (or increase) mesh density is when you're joining two objects, because a similar density at their common seam is needed.

Selecting Sculpt | Brushes opens a menu of volume and surface brushes that can drastically alter the model. We'll discuss them later in this chapter, but right now look at the Refine (add) and Reduce (subtract) icons at the bottom of the volume brushes page (see Figure 6-40). Press w to make the mesh appear, and drag the brush all over the

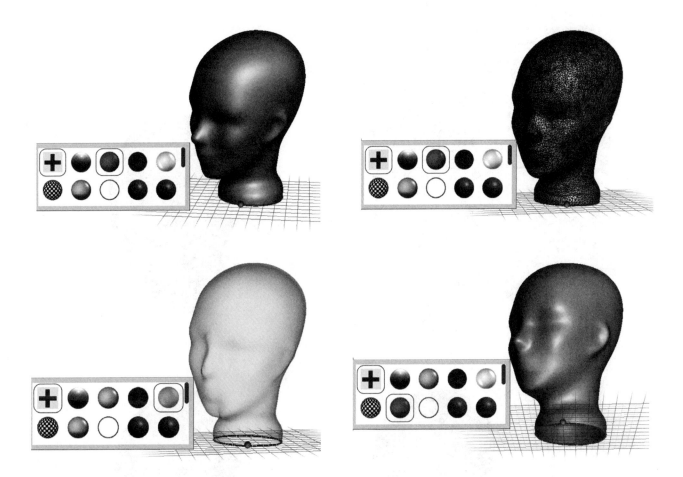

Figure 6-38 Here you can see the default shader, the default shader with visible mesh, the clear shader, and the red tip shader.

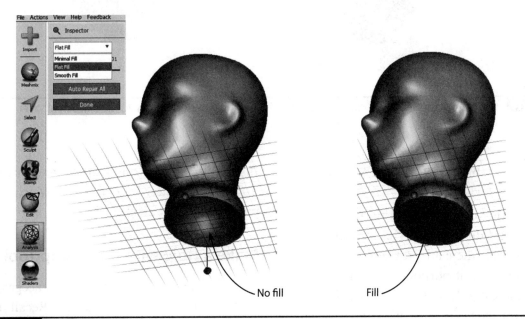

No fill

Fill

Figure 6-39 Use the Analysis | Inspector tool to fix the bottom hole.

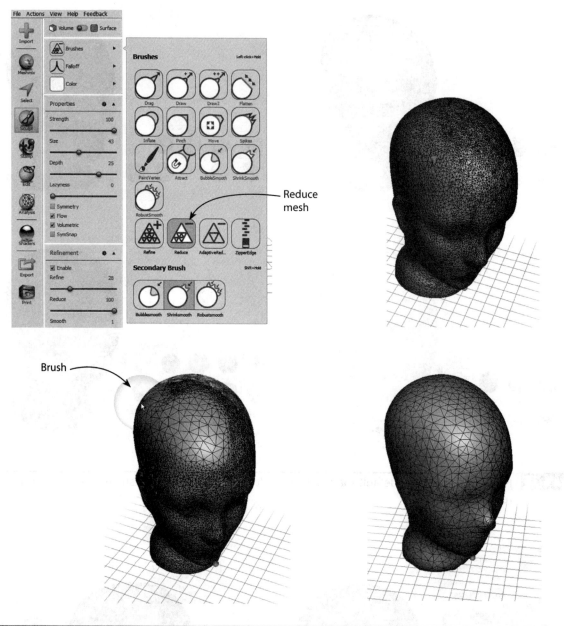

Figure 6-40 Reduce mesh density by clicking Reduce Mesh and then dragging a volume brush all over the model.

model. Click ESC when you're done, which saves the reduction and exits the tool.

Alternatively, you could select the whole head and click Edit | Reduce. This is quicker, but won't work if there are problems with the mesh. For instance, if portions of the mesh turn pink, that indicates the polygons are inside-out. Only the front side of a polygon can be selected. Also, there may be non-manifold edges. Select

Inspector | Analysis and fix the pinned areas. You can also select Edit | Remesh, which will remesh whatever is selected.

Make a Hole with the Stamp Tool

 The Stamp tool inserts a facegroup into the model's surface. Recall that a facegroup is a premade collection of polygons that can be operated on as a whole. A

stamp modifies the geometry of the surface to match the stamp shape. Follow these steps to make a hole with the Stamp tool:

1. Click a round stamp onto the glass head by clicking Stamp in the menu panel (see Figure 6-41). Activate the desired stamp by clicking its graphic and then left-click on the model in the location to place the stamp. While holding down the cursor, drag the stamp. Dynamically changing numbers on the right tells you the size. When you have the size you

want, let go. The stamp isn't immediately visible when you first click the model, but you should be able to see it while dragging. If you can't, change the model's face texture. The displayed numeric size makes using a stamp a way to select mesh more precisely. Check the Snap Dimension box at the bottom of the Stamp tool's screen for more accuracy.

2. Carefully run the selection brush over the polygons inside the stamp (zoom in to make this easier). You can see how much faster a

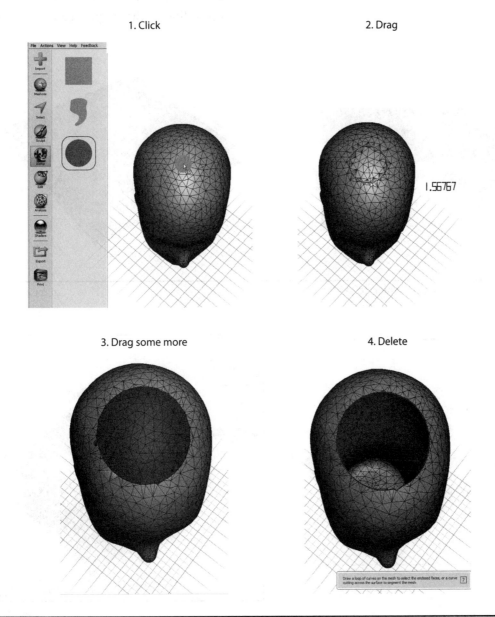

Figure 6-41 Drag a stamp onto the model, release the cursor, drag to the size wanted, and then delete it.

lower density mesh at the stamp's borders can be selected than a higher density one. Alternatively, you can paint a dab on the stamp with the Selection tool (the spot must not touch any mesh surrounding the stamp) and click Modify | Expand to Groups. When the stamp is fully selected, press the DELETE key. Done!

This worked because the model is hollow. Had you filled the head when healing the hole at the bottom, you'd need to click Edit | Extrude and push the stamp down to create an opening.

Make a Custom Stamp

If the stamps that come with Meshmixer aren't enough for your needs, you can make your own. Do this by sketching with the Select tool and converting the sketch into a stamp. Here's how:

1. Launch Meshmixer and import a plane (see Figure 6-42). We need it to sketch on.

2. Sketch a shape. Paint it with the Select tool and then click Smooth Boundary (see Figure 6-43).

3. Double-click anywhere on the boundary line to bring up an inflection point. Drag the point to finesse the shape (see Figure 6-44).

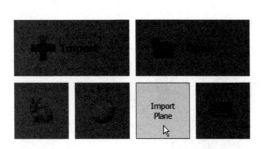

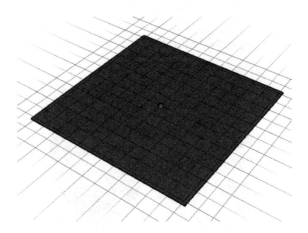

Figure 6-42 Import a plane on which to sketch the stamp.

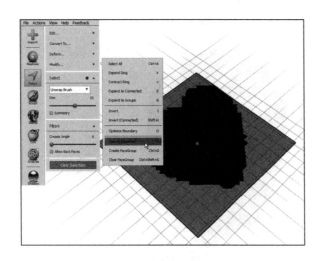

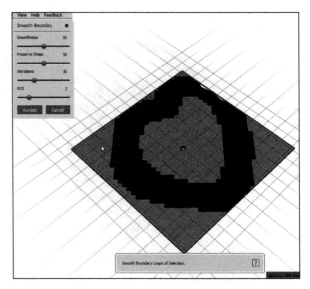

Figure 6-43 Paint a shape with the Select tool and smooth its boundary.

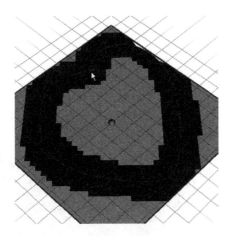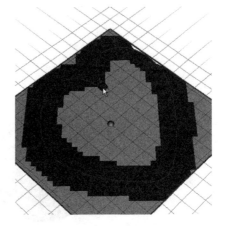

Figure 6-44   Double-click the boundary line to bring up an inflection point and then drag it to finesse the shape.

4. While still in Select mode, click Convert To... | Convert to Stamp (see Figure 6-45). The stamp will appear as an icon in the Stamp bucket and be there each time you launch Meshmixer. To remove it, right-click its icon and choose Discard Stamp.

Model a Bead

Let's model a round bead and stencil a face on it. Launch a new Meshmixer file and click the sphere. It will be the bead's body.

 Click on File | Import Plane. This brings up the same choices as the six square buttons of the launch screen. (Alternatively, click the plus sign icon on the menu panel to access a navigation browser to your own files). You'll be asked to append or replace the current file (see Figure 6-46). Append to add to it. Replacing would close the sphere and open the plane. Note that both models now appear in the Object Browser box.

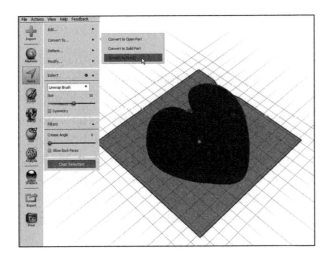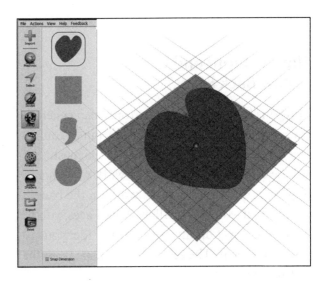

Figure 6-45   Convert the sketch to a stamp.

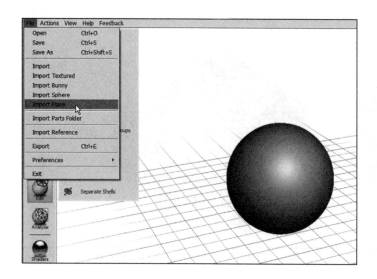

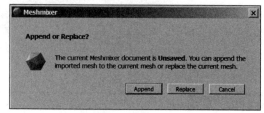

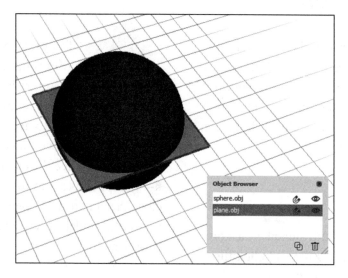

Figure 6-46   Import a sphere and append a plane to it.

Flip the Front and Use "Allow Back Faces"

Polygons have two sides: the front (also called the *normal*) and the back, which is pink. You can't select the back. Pink parts in the mesh will be a recurring issue as you work with this app. Selection brushes won't work on the back. However, if you check Allow Back Faces, you'll be able to select on the non-normal side; you just won't see it. The selection will appear on the normal side.

Our file will be easier to work on if the plane's normal faces away from the sphere. Because it doesn't, flip it 90° (see Figure 6-47). Highlight the plane by clicking it or its entry in the Object Browser box. Activate the Select tool, brush a dab on the plane's normal side, and click Modify | Expand to Connected to select all the plane. Then click Edit | Flip Normals. The result will be the plane's normal side facing away from the sphere.

Back side
(pink
colored)

Front/normal side

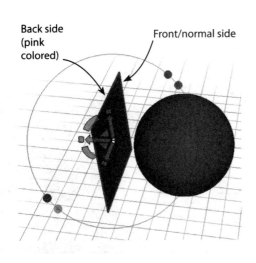

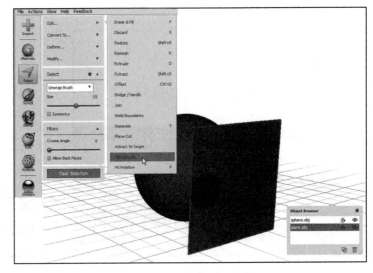

Figure 6-47 Rotate the plane and flip it so the front side faces away from the sphere. Check the Allow Back Faces option to select on the otherwise-unselectable back side.

Pierce the Sphere with an Extruded Stamp

We'll use a round stamp as a basis to create geometry needed to make a hole in the bead (see Figure 6-48).

1. Change the face texture and add the round stamp found in the Stamp menu. Drag the transparency slider on the model to make centering the stamp on the bead easier. Press w to see the mesh to make the stamp visible.

2. Select the stamp by brushing a dab on it and clicking Edit | Modify | Expand to Groups (see Figure 6-49). It may be hard to see with the transparent face; you can change its color by going to the menu bar and clicking View | Mesh Color Mode | Group Color.

3. Extrude the stamp by clicking Select | Edit | Extrude to bring up an options box (see Figure 6-50). At the top is the Offset slider. Because we placed the plane's front opposite

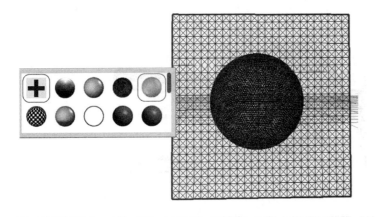

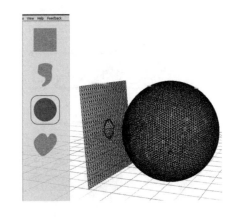

Figure 6-48 A clear face texture and visible mesh makes positioning a stamp on the sphere easier.

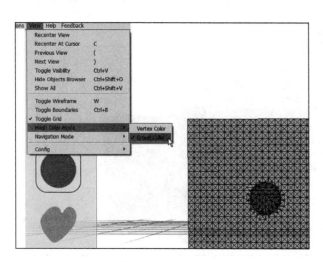

The colored stamp

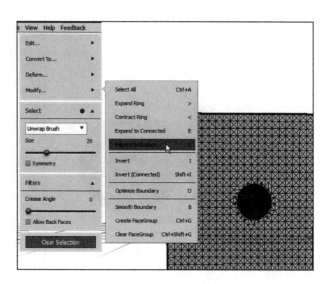

The selected stamp

Figure 6-49 Changing the stamp's group color may make it easier to see to select.

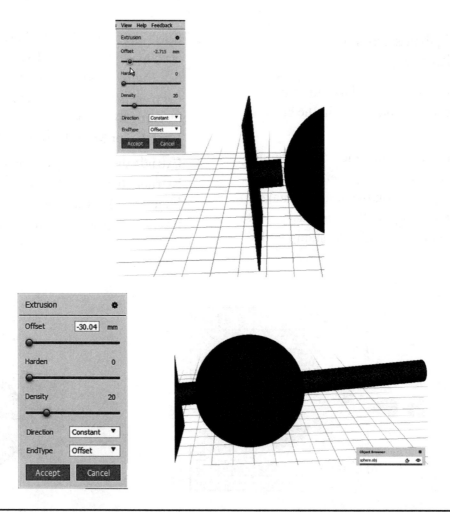

Figure 6-50 Using Select | Edit Extrude, give the stamp its length by clicking the slider number to bring up a text field and typing an appropriate number.

the sphere, move the slider to the left, which is the negative direction. However, even when the slider reaches the far left, the extrusion isn't long enough to reach the sphere. Click the offset number to make a text field appear. Type any number that will make an extrusion long enough to pierce the whole sphere and then click Accept.

Make the Bead Hole with Boolean Difference

To subtract the extruded stamp from the sphere, as shown in Figure 6-51, highlight both models in the Object Browser box (or click one model, hold down the SHIFT key, and then click the second model). Click the Menu panel's Edit icon. The options menu that appears will have different choices than when just one model is highlighted—specifically, the Boolean functions. Union, Difference, and Intersection appear, plus a function called Combine. Click Boolean Difference and then click Accept. Meshmixer detects what is inside and outside, and subtracts the extrusion from the sphere.

Boolean Intersection keeps two models' intersecting parts and discards the rest. Boolean Union joins two pieces together permanently. Before a model can be printed, all its pieces must be Boolean Unioned together. This isn't the same as Combine, which simply holds pieces together so they can be moved as one. Combined models can later be separated with the Separate Shells function (which is also in the options window), but Booleaned models are permanently altered. Boolean and Combine functions can only be done on two models at a time.

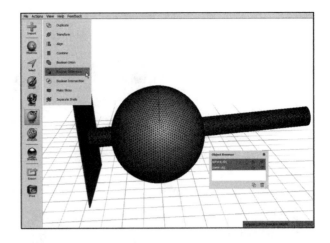

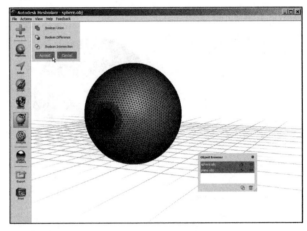

Figure 6-51   Apply Boolean Difference to subtract the extrusion from the sphere.

Add a Stencil to the Bead

A *stencil* is a 3D stamp. In the menu panel, click Sculpt and then click the top button to select Surface (Volume is the default). An icon for a quirky little feature called Stencils appears; click it. There are graphics for a few built-in ones, and a plus sign. The plus sign brings up a navigation browser where you can import your own .jpg, .png, or .xpm file. I imported the smiley face shown in Figure 6-52.

Click a graphic and then click it onto the model. If you don't see it, adjust its size and options, or try a different graphic. Some graphics work better than others as stencils.

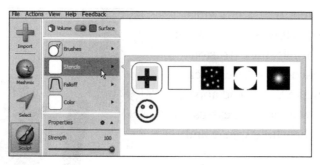

1. Import an image as a stencil.

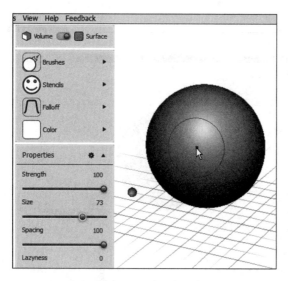

2. Apply the stencil to the sphere.

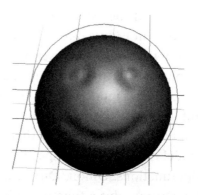

3. Finesse the stencil's size and options,
if needed, to make it display well.

Figure 6-52   Applying a stencil to the bead

Modify a 123D Design Pac-Man Model with Volume Brushes

In Chapter 3 we modeled Pac-Man and Pinky characters in Design (see Figure 6-53a). We're going to rough them up a bit here. You can export an .stl file directly from Design or download it from the Models section of the online 123Dapp.com account. Then import it via Meshmixer's launch screen (click the Import button and navigate to the .stl file). The model can also be exported directly from Design in its .123dx format by selecting it and clicking the Send to Meshmixer icon on the glyph that

appears when a selection is made (Figure 6-53b). Design will also let you select just a portion of a model to export, if that's what you want. You can select that portion, import it into Meshmixer, alter it, and export it back into Design at the same location.

In the menu panel, select Sculpt | Brushes. This brings up a selection of volume and surface brushes, with the button at the top indicating that Volume is selected (see Figure 6-54). Volume brushes create overlapping geometry. Surface brushes flatten the model. The brushes are labeled, but here are the ones that may not be self-explanatory:

(a)   (b)

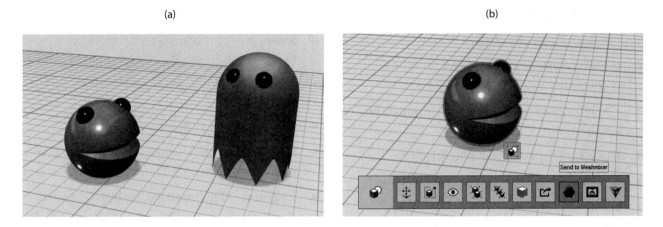

Figure 6-53   (a) Pac-Man and Pinky models made with 123D Design. (b) In Design, select a model or a portion of a model and then send it to Meshmixer by clicking the appropriate glyph icon.

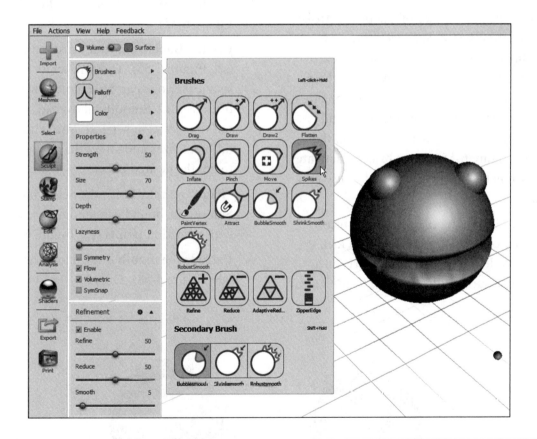

Figure 6-54   Volume brushes and their options

- **Drag**   Push the mesh around.

- **Flatten**   Lower the mesh's height.

- **Inflate**   Push the vertices (the polygons' points) out.

- **Pinch/Crease**   Pull the vertices toward the brush's center.

- **Draw 1**   Displace mesh and fill gaps.

- **Draw 2**   Displace mesh.

- **Paint**   Color the vertices.

- **Zipper**   Pull open mesh boundaries together.

- **Attract**   Pull selection toward a target object. You need two separate models in the workspace.

Falloff is the appearance of the brush stroke as it approaches the tip. *Secondary brushes* are modifiers (special effects) to the standard brushes; hold down the SHIFT key to use them. All brushes are used by clicking their icons to activate them (Spikes is activated here) and then clicking and dragging the cursor on the model (see Figure 6-55). Some brushes activate a white symmetry line on the model to use as a sculpting aid.

Unlike the selection brush, a volume brush works on a model's non-normal side. In fact, that's how modelers deal with pink parts on a mostly normal surface; they volume brush it away.

Here are the brush options:

- **Strength**   This controls how much each brush application affects the model (a setting of 100 means that depth or smoothness is added quickly).

- **Size**   This is the brush radius.

- **Depth**   This affects how far (deep) the brush marks go into the surface.

- **Lazyness**   This controls the brush's stroke style and direction accuracy (a setting of 0 means smooth flowing lines, 100 means straight ones).

We are now going to use two brushes and the Select tool to change Pac-Man's appearance.

1. Use the Spike brush on Pac-Man's head (see Figure 6-55).

2. Use the Bubble Smooth brush on the mouth (see Figure 6-56). This will remove the mouth so another one can be sculpted there. Then brush Shrink Smooth all over to smooth out the model in general. You'll need to try different sizes and strengths of these brushes to get the desired effect.

3. "Paint" a mouth with the Select tool. Smooth its boundaries and finesse the mouth with inflection points (see Figure 6-57).

4. Extrude the mouth inward and then delete it (see Figure 6-58). Done!

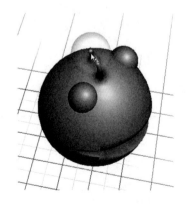 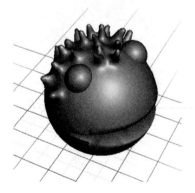

Figure 6-55   Use the Spike brush for this effect.

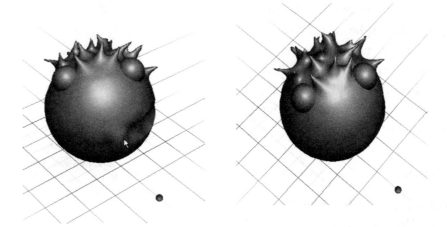

Figure 6-56 Use the Bubble Smooth and Shrink Smooth brushes to close the mouth and smooth the entire model.

Figure 6-57 Paint and finesse a mouth with the Select tool.

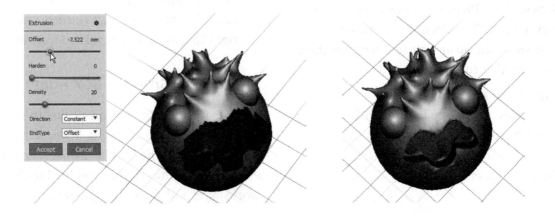

Figure 6-58 Extrude and delete the mouth.

Experiment! For instance, try out the Flatten brush on the eyes. Import spheres, scale them down, and Boolean Difference them to make concave eyes. Use different options and falloff types on the Spike brush and study the results. Plane cut and extrude the bottom for a flat base. Play with the Surface brushes.

Export the Model

When the model is finished, click File | Save. It will save as a .mix file. Then go to File | Export (or click the Export icon in the Menu panel) and export it into any of the choices shown in Figure 6-59, except Collada, which is a "go-between" file to import your work into other programs. We'll be importing Pac-Man into another Meshmixer file later, and only .stl (both ASCII and binary), .obj, .ply, and .amf will import. You can open a .mix file in Meshmixer, but you cannot import one.

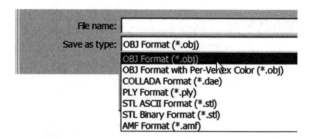

Figure 6-59 Export the .mix file as an .obj, .stl, .ply, or .amf.

Duplicate and Modify Pinky

Time to open Pinky's .stl file. We'll copy him twice and add arms from the Meshmix Parts Bucket. Here are the steps to follow:

1. Select Edit | Duplicate (see Figure 6-60). The model will become dark because it's no longer active. A duplicate has been made over it; you can see it listed in the Object Browser box. Press T and move the duplicate over with the Transform tool. Then make another duplicate and move it over, for three Pinky models total.

2. Import and append a plane via File | Import to serve as a base (see Figure 6-61). Note that it has an entry in the Object Browser box, too. Move it into place with the Transform tool.

3. Extrude the plane (see Figure 6-62). We want to extrude its underside. However, that's the non-normal (pink) side, which can't be selected. Check the Allow Back Faces box. Now we can brush a dab on the underside, click Select All, and the whole plane will be selected, as evidenced from its top view. Click Edit | Extrude and move the slider to the desired depth. Move the Harden slider to the right to make the edges sharp instead of soft.

The Object Browser Box Again

Let's revisit the Object Browser box to note its relationship with the models. When a model is highlighted in the Object Browser box, the corresponding model is highlighted in the workspace, and shows up lighter than everything

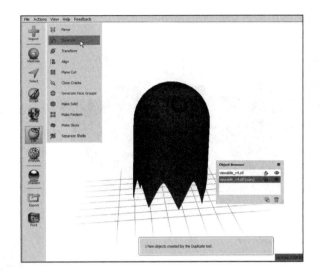

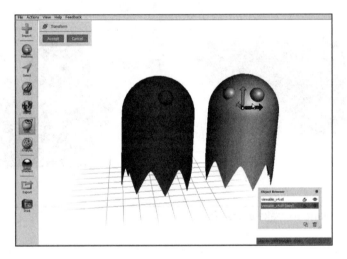

Figure 6-60   Duplicates appear over the original. Move them aside with the Transform tool.

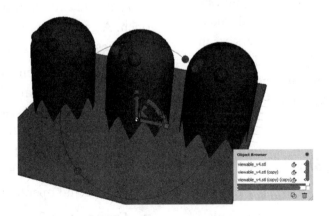

Figure 6-61   Import and append a plane to serve as a base.

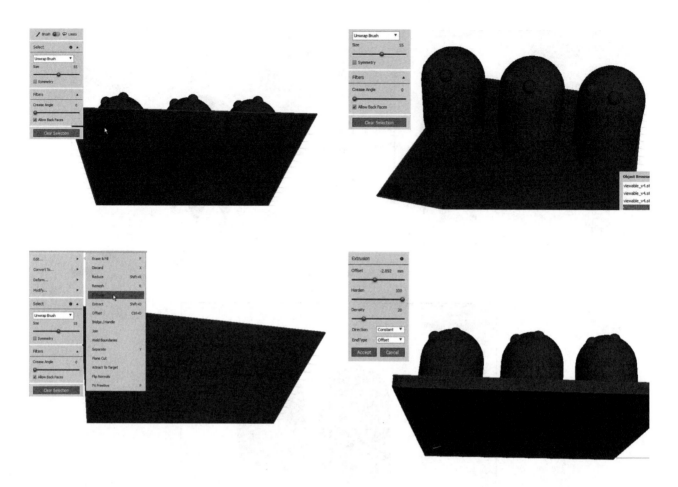

Figure 6-62   Extrude the plane to make a base for the models.

else (see Figure 6-63). You can highlight a model by clicking it directly or clicking on its entry in the box. Remember, too, that you can hide models via the box, which is useful when you want to work on one and the others are in the way. You can delete a highlighted model by clicking the box's trash can icon, and you can duplicate a model by clicking the box's double-square icon.

The Meshmix Parts Bin

The Meshmix Parts Bin is a fun feature that lets you add pre-made parts to the model. You can also make and save your own parts. Click the menu panel's Meshmix icon. A Primitives library and flyout

arrow to access more libraries appear (see Figure 6-64).

There are two kinds of parts: solid and open. A graphic in the part's lower-right corner identifies which it is (see Figure 6-65).

- **Solid part**   When you drag a solid part onto the model, it retains its independence. For instance, you can select-brush a dab on it, choose Edit | Modify | Expand to Connected or Edit | Modify | Expand to Groups, and only that part will be selected. It can be easily deleted and operated on.

- **Open part**   When you drag an open part onto the model, that part becomes integral to it; think of it as a Boolean Union. You can't select or otherwise operate on it independently.

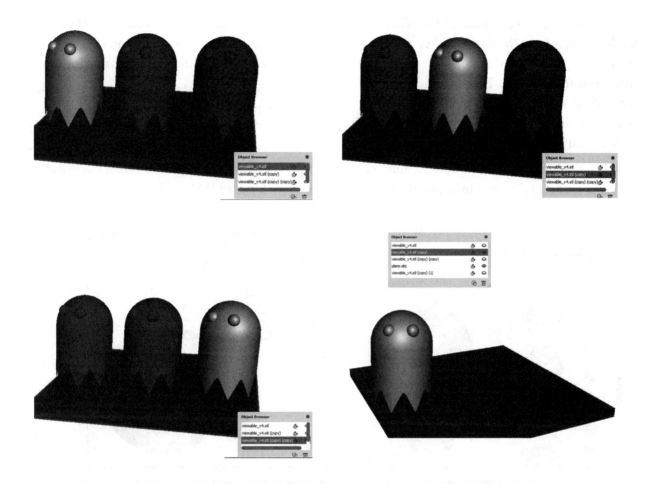

Figure 6-63   Models can be highlighted, hidden, and duplicated through the Object Browser box. Highlighted models show up brighter than everything else. In the last graphic, two of the models are hidden.

Figure 6-64   The Meshmix icon accesses parts libraries

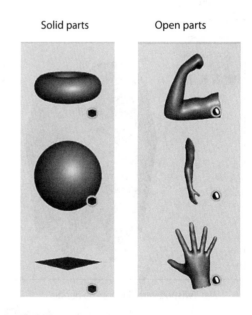

Figure 6-65   Parts are open or solid, and identified as such by a graphic in their lower-right corner.

You can turn an open part into a solid part by dragging it from the Parts Bucket into the workspace. When you drag an open or solid part into the workspace, the Transform tool appears for scaling, rotating, and moving. You can operate on it there, move it to the model, finesse it, and easily change it some more later.

When you initially drag an open or solid part onto the model, a different manipulator widget appears (see Figure 6-66). You can scale, rotate, move, and twist with it. Once you click Accept, future tweaks are done with the Transform tool.

Turn any model or portion of it into a solid or open part by selecting it and clicking Edit | Convert To.... It will automatically be added to Meshmix | My Parts, where you'll click to access it. To delete it, select its icon, right-click, and choose Discard Part.

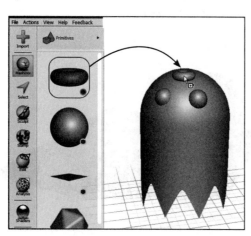

Drag a part onto the model.

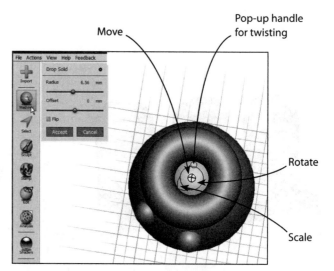

A widget appears when the part is placed on the model. The part can be manipulated with that widget.

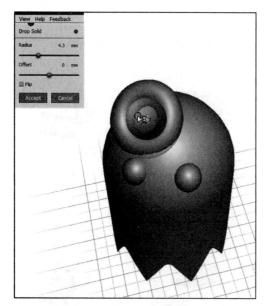

Twisting with the pop-up handle

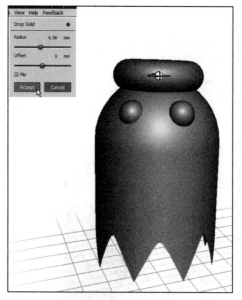

Click Accept when finished.

Figure 6-66 When a part is dragged from the Parts Bin onto the model, a manipulator widget appears. Make changes to the part on the widget or in the options box.

Add a Parts Bin Arm to Pinky

Drag an arm from the Meshmix Arms Bucket onto the workspace. We need it to be a solid part because we'll be operating separately on it after it's placed on the model. Also, when you put an open part on the model, the manipulator widget tends to warp the surrounding area.

If you can't find a part after you drag it into the workspace, it may be much larger than the model; click the Focus icon on the Hotbox to bring everything into sight. Then use the Transform tool to scale and move the part into place (see Figure 6-67). The arm is brighter than Pinky because it's a separate model and is the active model. If you have difficulty selecting it, click its entry in the Object Browser box.

Mirror the Parts Bin Arm

Any model can be mirrored, which creates a duplicate that is positioned opposite it. Only whole models can be mirrored, not selected parts of them. The original and the mirror are welded together, but can be separated.

Select the arm and then click on Edit | Mirror. A mirroring plane and the Transform tool appear; rotate the mirroring plane into place. Don't be alarmed if the mirrored arm disappears—it will reappear when the plane is correctly positioned. By default, the plane appears at the origin, so if the model is off that, the plane will have to be manipulated. In the first graphic of Figure 6-68, you can see a bit

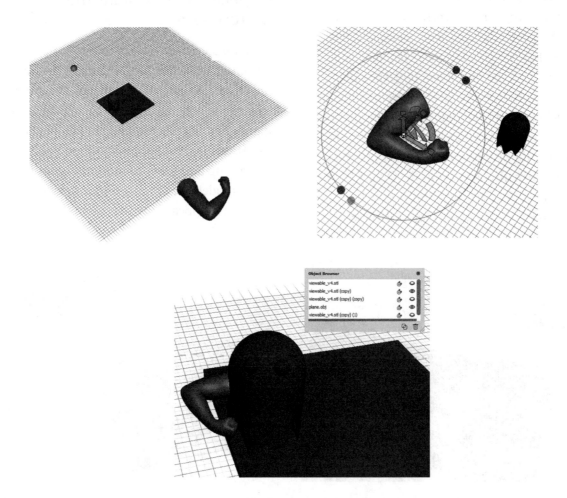

Figure 6-67 Drag an arm into the workspace, scale it, and move it in place.

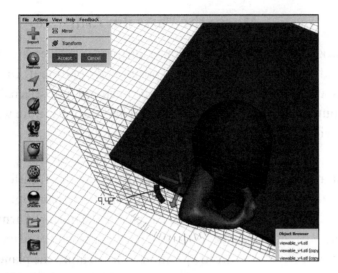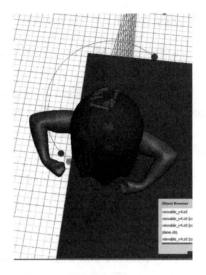

Figure 6-68 Click Edit | Mirror, rotate the mirroring plane, and move the new arm into place.

of the second arm. Move the plane to the right and click Accept. The mirrored arm will turn green. You can select the arms separately, but because they're welded to the original, if you move one with the Transform tool, the other will move with it. Solve this by selecting one arm and clicking Edit | Separate Shells. This makes them independent, which is needed to copy each arm to the Pinky duplicates. An entry for each duplicate arm also appears in the Object Browser box. Unhide the other two Pinkys now by clicking the eye icon in the Object Browser box.

Duplicate the Original and Mirrored Arms

We have a couple of ways to duplicate and move the arms. One is to highlight the model in the Object Browser box and click the duplicate icon in the box. Another is to click the arm to highlight it and then click Edit | Duplicate. If the Transform tool doesn't appear, press T to bring it up and use it to move the arms to the other models (see Figure 6-69). Remember that duplicates appear over the originals.

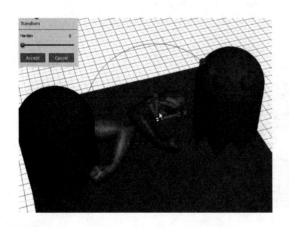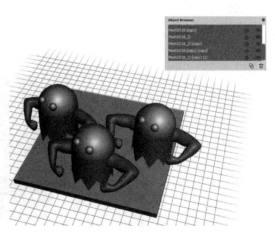

Figure 6-69 Duplicate the arms and move them to the other Pinky models.

We need a bigger base to make room for Pac-Man, whom we'll import shortly. There are a couple of ways to do this:

- *You can duplicate, move, and scale the base.* Highlight the base by clicking it or its entry in the Object Browser box. Duplicate it, move the duplicate forward, adjust its length with the Transform tool, and move it into place (see Figure 6-70).

- *You can scale the existing the base larger.* Highlight the base by clicking it or its entry in the Object Browser box. Press T and drag the Transform tool's red scaling box to make the base longer (see Figure 6-71). The base will scale from the center. You can use Plane Cut to trim excess.

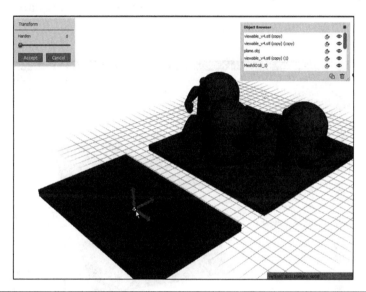

Figure 6-70 Enlarge the base by duplicating, moving, and scaling it.

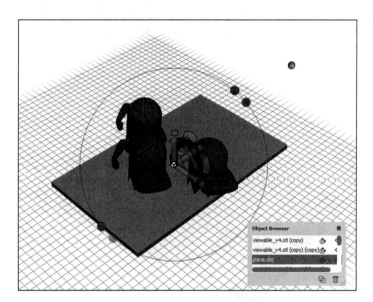

Figure 6-71 Make the base longer by dragging the Transform tool's red scaling box. The base will scale from the center.

Combine the Pinky Models with Their Arms

Scaling the base larger requires moving the Pinky characters back a bit to make room for Pac-Man. Because each Pinky and each arm is a separate model, all nine must be highlighted (click each while holding down the SHIFT key) to move them together. When moving multiple models a lot, it's most efficient to combine them. Highlight one Pinky and one of its arms, then

select Combine from the menu that appears (see Figure 6-72). This holds them together so they can move together; they can always be separated via Separate Shells. Only two models can be combined at a time, so repeat the process with Pinky and his other arm.

Finally, import Pac-Man (see Figure 6-73). Select File | Import, navigate to his .stl file, and bring him in. He'll need to be scaled and moved. Before placing him, make a plane cut at the bottom so he'll sit squarely on the base.

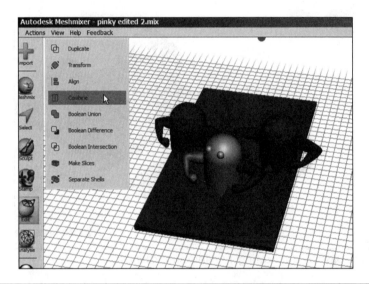

Figure 6-72 Combining models enables them to be moved together. Only two can be combined at a time. Here, the front Pinky and his right arm are highlighted for combining.

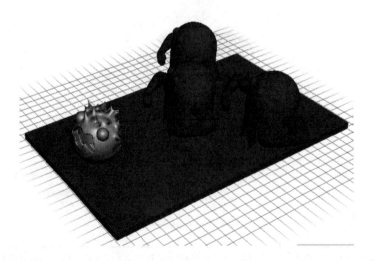

Figure 6-73 Import, plane cut, and place Pac-Man.

This should get you started with Meshmixer. I say "started" because there are other fun features. Check out the Pattern tool, which modifies objects with different pattern techniques, and the Slice tool, which cuts one model into multiple models. The Soft Transform function deforms the model in a nonlinear manner (it treats the selection boundaries like rubber). Experiment with the tools' option sliders to see how different settings affect the result. Download files from the 123D Gallery and Thingiverse and then meshmix them up. Definitely read the Meshmixer blog and support forum (links are provided at the end of the chapter) to keep up on the latest developments.

One critical function we haven't covered in this chapter is how to use Meshmixer to prep a file for 3D printing. Because that process applies to files made with all the 123D apps, it is discussed in Chapter 9. Therefore, if you're in a hurry to print what you meshmixed, head on over there.

Summary

We used 123D Meshmixer to alter and combine .obj and .stl models and make models from scratch. You saw how to navigate the interface, select mesh, use brushes, heal holes, and perform basic editing with the Transform, Plane

Cut, Boolean Difference, and Mirror tools. Meshmixer's great value is that it enables you to edit a solid or mesh model made in any program as along as you have it in an .stl, .obj, .ply, or .amf format.

Sites to Check Out

- **Meshmixer Blog**   blog.123dapp.com/category/meshmixer.

- **Meshmixer support forum**   www.meshmixer.com. Check the announcements board for updates and a comprehensive list of hotkeys.

- **Meshmixer YouTube channel**   www.youtube.com/user/meshmixer.

- **Download models from the Smithsonian to meshmix**   3d.si.edu/browser.

- **Meshmix a Ghostapus**   www.instructables.com/id/Ghostapus/step5/Big-Octopus.

- **Free software for processing mesh models**   meshlab.sourceforge.net.

- **Free software for processing .stl files**   www.netfabb.com.

- **Information about the .amf file format**   www.shapeways.com/blog/archives/898-amf-a-better-file-format-for-3d-printing.html.

- **Meshmixer on Twitter: @meshmixer**

CHAPTER 7

For the Kids! Tinkercad

TINKERCAD IS A WEB-BASED program written to introduce elementary school kids, or anyone with little computer experience, to solid modeling and 3D printing. Therefore, it is simpler than 123D Design, doesn't require as much computing horsepower, and works best on small projects. Yet it has some powerful features, such as the ability to import .stl and .svg files, export .stl and Minecraft-ready files, and make unique forms with a feature called the Shape Generator.

Because it is also free with an affordable subscription option, it's very popular in K-12 education settings. Both free and paid accounts can store an unlimited number of files. Check out what other tinkerers are making at www.tinkercad.com/things/ (see Figure 7-1).

In this chapter, you'll learn how Tinkercad is used via projects that incorporate the features just discussed. Tinkercad is accessible from almost any device, across the Windows, Mac,

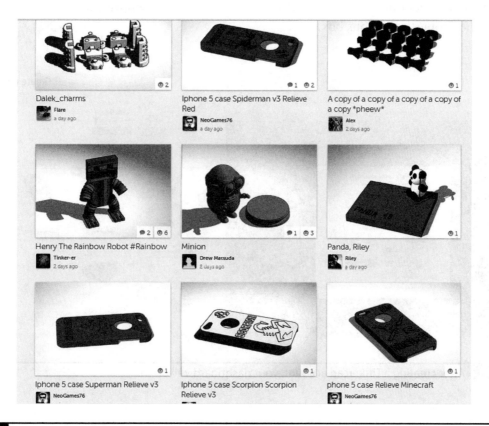

Figure 7-1 A page of models at www.tinkercad.com/things/

and Linux platforms. Your browser must be HTML5/WebGL-enabled, and Chrome works best. Point it to www.tinkercad.com. You'll be asked to make an account; it's separate from the account used for the rest of the 123D suite. Once you make that account, your dashboard will open. Because you don't have any projects, it will mostly display tutorial projects, as shown in Figure 7-2. We'll discuss the dashboard in depth later; right now, just click the Create New Design button, and let's tinker!

The Tinkercad Interface

The Tinkercad interface, shown in Figure 7-3, consists of the Tinkercad graphic, a menu bar, navigation wheel, fit and zoom icons, grid adjustment settings, an arrow that exposes a library panel, and a workplane. Here's a description of the items you'll find there:

- **Generated filename**   Tinkercad assigns a random name to your file. You can rename it through the Properties options box.

- **Menu bar**   The menu bar contains the following icons:

- **Tinkercad graphic**   Click this to go to your dashboard.

  - **Design**   Contains utility functions, Properties options, download links, upload links to third-party service bureaus, and links to the Thingiverse and a gallery of shared models.

  - **Edit**   Contains the Copy, Paste, Duplicate, and Delete functions for parts within the file.

  - **Help**   Contains links to tutorial videos and other information.

  - **Undo**   Undoes each step, one at a time.

  - **Redo**   Redoes each step, one at a time.

  - **Adjust**   Contains the Align and Mirror functions. Align lets you select-snap the positions of multiple parts so their edges or centers line up. Mirror lets you reverse a part's position.

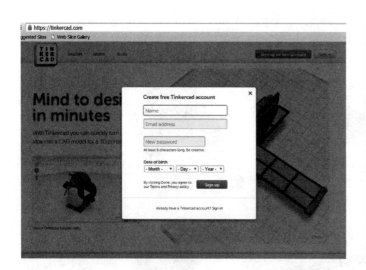

Figure 7-2   After creating a Tinkercad account, your dashboard will open.

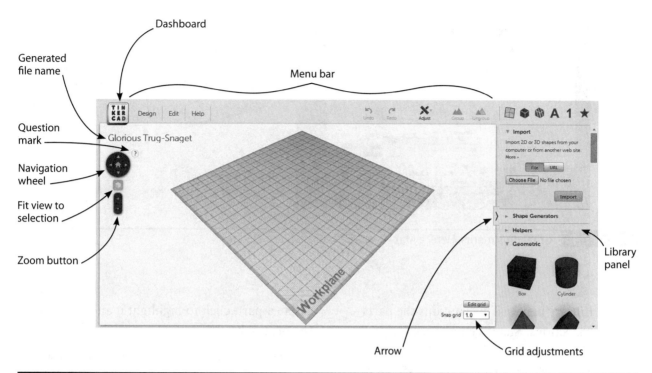

Figure 7-3   The Tinkercad interface

- **Group**   Holds selected parts of the model together so they can be moved together.

- **Ungroup**   Returns selected grouped parts back to individual parts. Ungroup may need to be clicked multiple times to ungroup nested groups (groups inside larger groups).

- **Navigation wheel**   Click the house icon to return the workplane to the default position. Click the arrows to move the workplane in those directions.

- **Fit View to Selection**   Select something on the model and click this item. The select part will fill the screen. If nothing is selected, the whole model will fill the screen.

- **Zoom buttons**   The plus sign gives you a closer view of the model; the minus sign gives you a farther-away view.

- **Mouse**   The mouse is your other navigation tool. Rotate the scroll wheel to zoom in and out; hold it down to pan (slide the model around the screen); hold down the right button to turn (which orbits you around the model); hold the left button to move (drag) the model into a different location on the workplane.

- **Question mark**   Click this item for a pop-up chart of navigation keys.

- **Grid adjustments**   Snap Grid aligns the model with the nearest intersection, in increments of your choosing. Set it to Off to move the model freely. Edit Grid lets you choose units, the width and height of the squares, and a preset for a specific 3D printer (see Figure 7-4).

- **Arrow**   This opens and closes the Library panel.

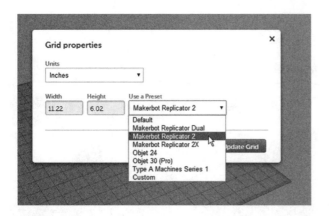 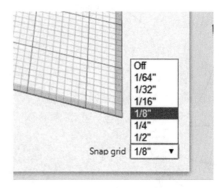

Figure 7-4   Grid adjustments. Here, a Makerbot and 1/8" snap are selected.

- **Library panel**   This contains the parts collections: Shape Generators, Helpers, Geometric, Holes, Letters, Number, Symbols, and Extras. Figure 7-5a shows the collapsed libraries; click the icons at the top to quickly access a specific collection, handy when most of the collections are open. Figure 7-5b shows the specific collections.

- **Workplane**   This is the flat, gridded surface on which all modeling is done. Unlike the workplane in other apps, it has no origin. Parts snap to workplanes, not to each other.

Tinker Basics: Drag, Delete, Select, Move, and Save

Models are built by dragging parts (click a part and hold down the left mouse button) from the Library panel and altering them. Press a keyboard arrow key to move an item 1 mm. Abort a dragging operation with the ESC key. Otherwise, the ESC key has limited functionality here, so typically you'll complete an operation and click Undo if you don't want to keep it. To

delete a part, click to highlight it and then press the DELETE key.

The SHIFT key serves several purposes. When moving a part, hold the SHIFT key down to keep the part aligned with the x or y axis. When using the arrow key to move a part, press SHIFT to move it 10 cm. When scaling a part, hold the SHIFT key down to keep the shape proportional along all axes. And when rotating a part, hold the SHIFT key down to snap the rotation at 45° increments.

Select a part by clicking it. Select multiple parts by clicking and holding down the SHIFT key. Alternatively, select multiple parts by dragging a crossing window around them with the mouse (click the left button, drag, and release). This window will select everything it touches, regardless of whether the part is entirely inside the window. A crossing window needs to be started on a workplane, not on a part. Clicking and dragging a part moves that part on the workplane.

Tinkercad saves your work automatically. It remembers everything you do; hence, it can undo steps even after you close and reopen the file. However, once you click Design | Save, you cannot *redo* past the point of that save. Future redos will affect changes made after the save.

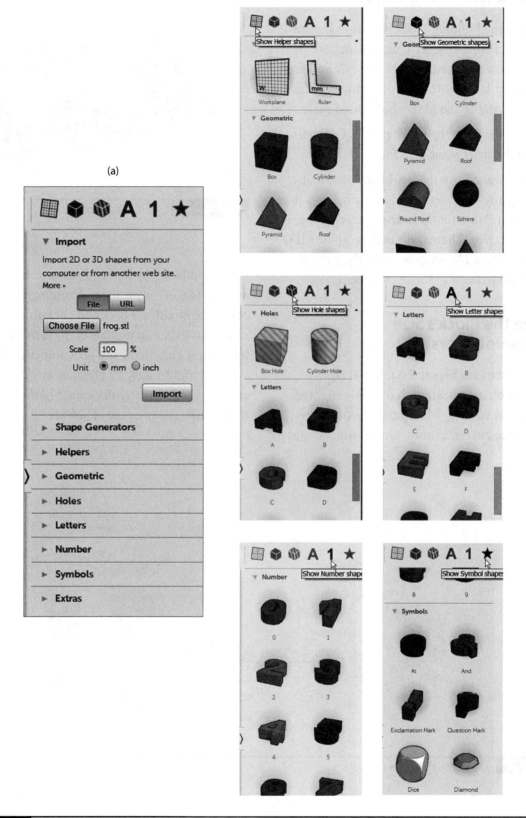

Figure 7-5 (a) The collapsed Libraries panel. (b) The Library panel's collections.

Tinker a Building Block

Let's tinker the building block shown in Figure 7-6. In Grid Properties, set Units to Inches, select the preset Makerbot Replicator 2, and set Snap Grid to 1/8". Note the workplane now has the proportions of the Makerbot's build plate.

Drag a cube from the menu panel's Geometric library to the workplane (see Figure 7-7).

Once the cube is on the workplane, move it by clicking anywhere on it and dragging it to the new location. Note that it is displayed as a three-point perspective. There is no orthographic (2D) mode. Because a 1/8" snap was set, the cube will snap to intersections at those increments.

Change the Block's Size with Manipulators

Click the cube (see Figure 7-8). It will be outlined in blue, indicating that it's highlighted and ready for editing. Manipulator icons also appear: three curved arrows, an arrowhead, and

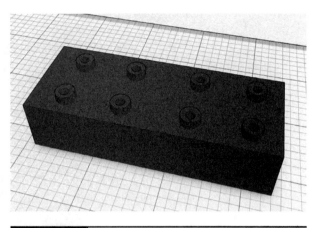

Figure 7-6   A building block

white and black grips. An Inspector box appears as well. It contains options to change the part's color (highlight the part and click the color swatch to bring up a palette), to turn the part into a hole (highlight the part and click the hole swatch), and to lock all changes made so they can't be inadvertently undone. Figure 7-9 shows the manipulators in use. They are:

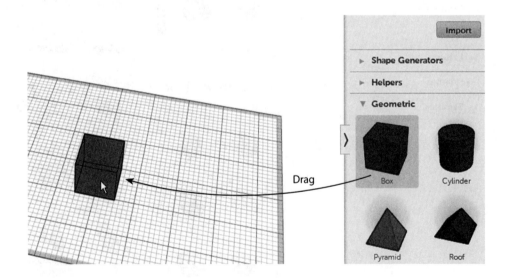

Figure 7-7   Drag a cube to the workplane.

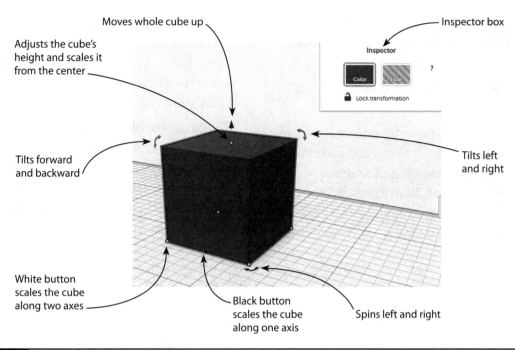

Figure 7-8 Click the cube to highlight it. This makes a thin blue outline, manipulator icons, and the Inspector box appear.

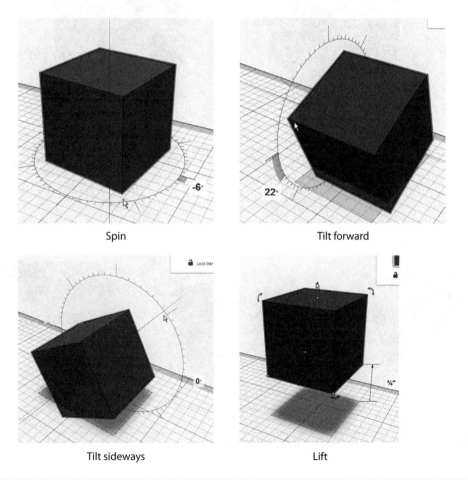

Figure 7-9 Move the cube with the manipulators.

- **Curved arrows** The curved arrows rotate and tilt the cube. The active arrow is black; the inactive ones are gray.

- **Arrowhead** The arrowhead moves the whole cube up and down.

- **Grips** The white grip in the middle is used to drag the block taller (drag up) and scale the block proportionately around its center (drag left or right and hold down the SHIFT key). The other white grips scale along two axes, but not around the center. Again, hold down

the SHIFT key to scale proportionally when dragging them; otherwise, the proportions will change. The black grips scale along one axis. Hover the mouse over any grip to see the part's dimensions (see Figure 7-10).

To change the block's size, grab the lower-left white grip and drag that side left until the block is 2-1/2" long (see Figure 7-11). Grab the upper-right white grip and drag that side until the block is 1-1/4" deep. Then grab the top-middle white grip and drag it up until the block is 3/4" high.

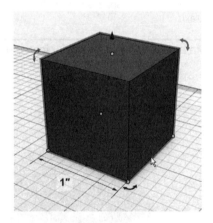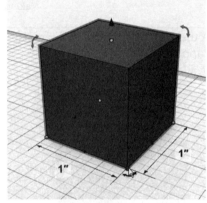

Figure 7-10 Click a white grip to see two dimensions; click a black button to see one dimension.

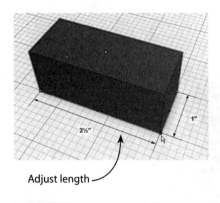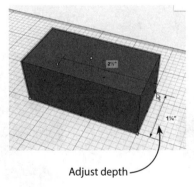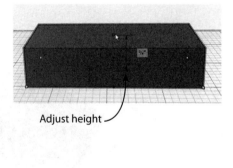

Adjust length

Adjust depth

Adjust height

Figure 7-11 Scale the block by dragging the white grips.

Snap Parts Together with the Workplane

All parts must go on a workplane. Although placing parts together without putting them first on a workplane is possible, it's much harder. So drag a workplane from the Helper collection onto a part (it can go on a straight, angled, or round surface), and then drag any other Library part to that workplane. The part will snap to the plane. Parts already in the model can be snapped together with workplanes, too. Drag a workplane onto one part, and then drag another part to that workplane. All dimension numbers that appear on a part are relative to the plane it is on or snapped to.

Once a workplane is clicked onto a part, it becomes the active workplane. Helper workplanes are orange to distinguish them from the original blue workplane. To reactivate the original workplane, drag a new Helper workplane onto it.

Just to see how this works, drag a new workplane from the Helper collection onto the default workplane to make it active again (see Figure 7-12).

Make the Studs with the Cylinder Part

Here's how to use a workplane and cylinder to make the block's studs.

Bring in a Helper workplane on which to build the studs by clicking Helpers | Workplane and dragging the plane onto the top of the block (see Figure 7-13).

Drag a cylinder onto the block by clicking the Geometric | Cylinder collection and dragging

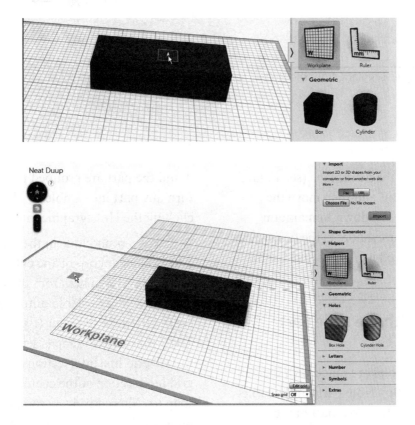

Figure 7-12   Make the default workplane active by dragging a Helper workplane onto it.

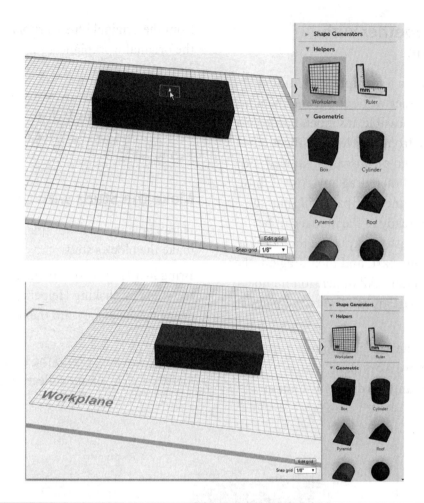

Figure 7-13 Drag a workplane from the Helper collection onto the top of the block.

a cylinder on top of the workplane (see Figure 7-14). Click to select it, and then move the grips (hold the SHIFT key down to maintain proportions) until the cylinder is the desired size. For this project, I sized it 1/8" tall and 1/4" in diameter.

Make a Hole with the Hole Part and Grouping

Round and square holes are made with the Cylinder Hole and Box Hole parts in the Hole collection. Drag a hole onto a part in the workspace. The hole will remain separate until it and the part are grouped together. You can turn any part into a hole by clicking it and then clicking the Hole graphic in the Inspector box.

To cut a round hole in the cylinder, first drag a workplane on top of the cylinder (left graphic in Figure 7-15). Then from the Holes collection, drag the Cylinder Hole onto that workplane, centered on the cylinder (right graphic in Figure 7-15). Since a 1/8" snap is set, the hole will snap to 1/8" grid line intersections. There should be a grid intersection at the center of the cylinder to snap to. Alternatively, if you just want to freely move the hole part, set the Snap Grid to Off.

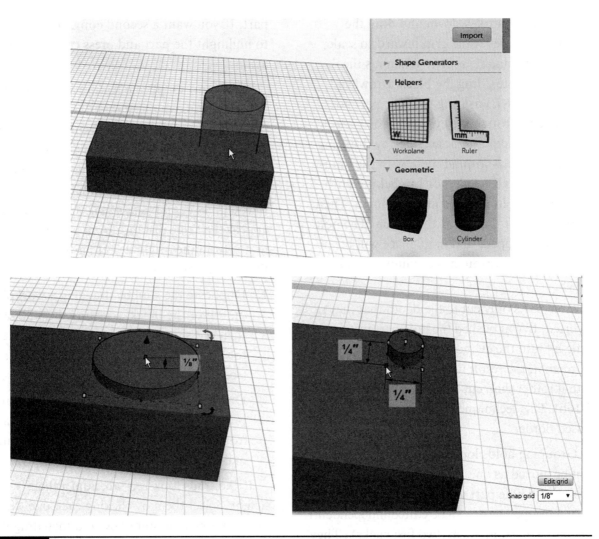

Figure 7-14   Drag a cylinder onto the block and use the grips to make it 1/8" tall and 1/4" in diameter.

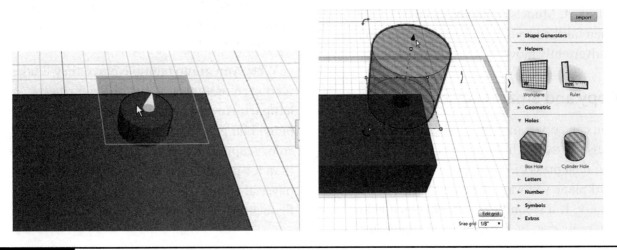

Figure 7-15   Drag a workplane onto the block and then drag a hole part onto it.

Duplicate feature a try:

Hold the SHIFT key down and drag the hole part's center white grip inward to scale it smaller around its center. When it's the desired diameter, drag the black arrowhead to make it taller, ensuring the hole part goes completely through the cylinder. Then select both the cylinder and hole either by dragging a crossing window around them with the mouse (click, drag, and release) or by clicking both parts while holding the SHIFT key down. Then click Group at the top of the screen. The hole part will disappear and in its place will be an actual hole cut through the cylinder (see Figure 7-16a).

Hole parts only cut to the depth that they're inserted. If you pushed the hole part just a little bit into the cylinder instead of all the way through it, the result would be a hole burned into it rather than cut through it (see Figure 7-16b). This technique is how you could make a raised border.

Grouping holds loose parts together so they can be moved as one. And as you just saw, grouping a hole part with another part cuts a hole. Grouped parts can be edited independently by double-clicking them (two fast clicks) . They can be ungrouped any time by highlighting them and clicking Ungroup. Ungrouping the cylinder and hole will turn the hole back into a hole part. Models should be grouped together when finished so that individual parts won't get inadvertently moved.

Copy and Smart Duplicate

You can copy a part by highlighting it and pressing CTRL-C (hold the CTRL key down and then press C). Then, click where on the workplane you want the copy to go and press CTRL-V. If a second part is highlighted before you press CTRL-V, the copy will snap to that part. If you want a second copy, you'll need to highlight the part and press CTRL-C/CTRL-V again.

The Smart Duplicate feature automatically makes multiple duplicates and remembers their offset distances. To use this feature, select a part and press CTRL-D. This duplicates it and puts the duplicate over the original. Drag the duplicate off the original and place it where you want it. If you press CTRL-D again, a second copy will appear the same distance from the first as the first is from the original. Let's give the Smart Duplicate feature a try:

1. Duplicate the stud by dragging a workplane on top of the block, selecting the stud, and then pressing CTRL-D. Drag the duplicate off the original and place it where you want it (see Figure 7-17).

2. Duplicate both studs by selecting the pair, pressing CTRL-D, and moving the duplicates. You can then select and duplicate all four to make eight studs total (see Figure 7-18).

If the studs look too tall, adjust their height. Select them by clicking each one while holding down the SHIFT key, and then drag the manipulator arrow down until the studs are the desired height (see Figure 7-19). Finally, group the block and studs together by dragging a crossing window around them and clicking Group. This enables you to move all the parts as one unit. Done!

When you group multiple parts, they'll take on one color, typically that of the farthest-left part. However, when you double-click individual parts to edit them, their original color reappears. Colors don't have any use other than for presentation purposes, because the physical model will be the color of the printer filament.

(a)

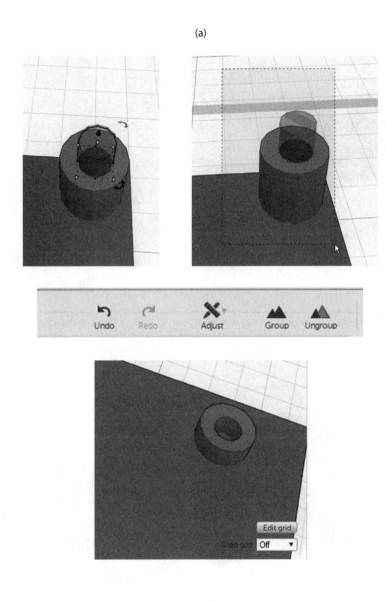

(b)

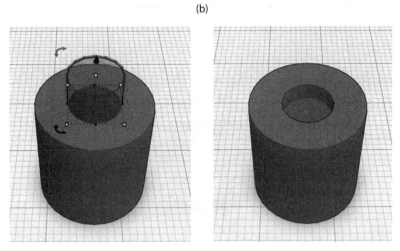

Figure 7-16   (a) Scale the hole part, select it and the cylinder, and group them together to cut a hole through the cylinder. (b) Hole parts only cut to the depth they're inserted.

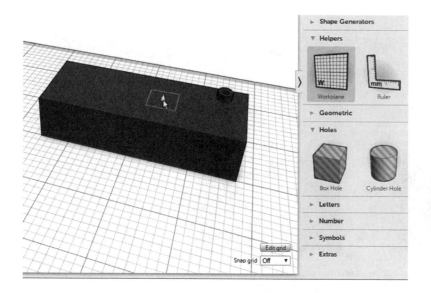

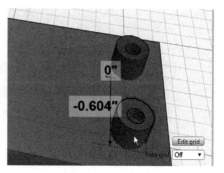

Figure 7-17   Duplicate the stud by highlighting it, typing CTRL-D, and dragging the duplicate off the original.

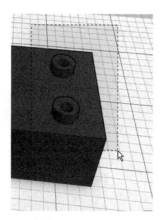

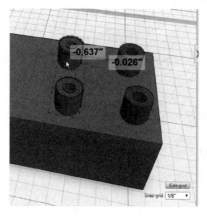

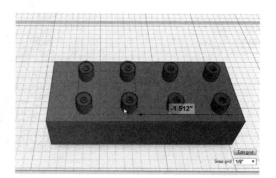

Figure 7-18   Duplicate the studs until there are eight total.

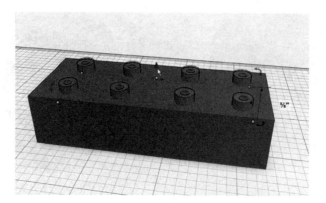

Figure 7-19   Make the studs shorter, if needed, by dragging the arrowhead down..

The Ruler

The ruler is a tool that measures along two axes. It verifies existing sizes and distances and can be used to change a part's size and location. Let's demonstrate how it works on the block. Follow these steps:

1. Drag the ruler from the Helper menu and place it at the block's lower corner (see Figure 7-20). If you want to place it on a top corner, drag a workplane on top of the block first and then drag the ruler to it. The ruler defaults to the upper-right quadrant, but clicking the circle cycles it through the other three quadrants. You can also drag the circle to move the ruler anywhere on the workplane. The ruler will snap to the grid unless the grid is set to Off.

2. Click anywhere on the block and the following ruler manipulators will appear: an X, dimension lines on all three axes, a black rotator arrow, and two arrowheads (see Figure 7-21a). Click the X to exit the ruler. Click the arrowheads to toggle between *midpoint* and *corner*. *Midpoint* adds a dimension line from one end of the block to the center, and *corner* shows a dimension from one end of the block to the other (see

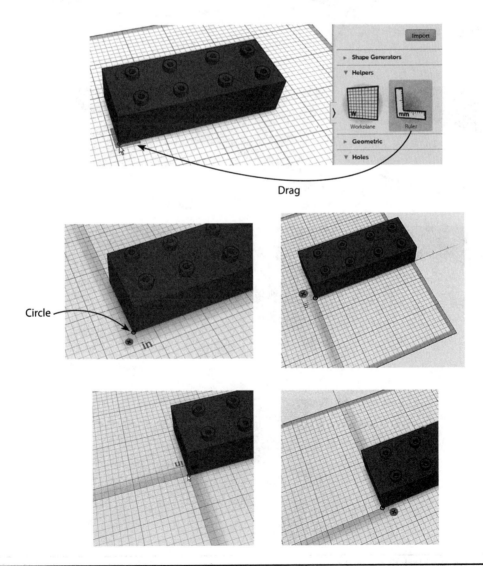

Drag

Circle

Figure 7-20 Drag the ruler to the block's lower corner. Change its quadrant location by clicking the circle.

Figure 5-21b). Click the curved arrow to rotate the block (see Figure 5-21c); hold down the SHIFT key to rotate it in 45° increments.

3. The x,y coordinates (0,0) that appear near the rotator arrow mark the ruler's origin (see Figure 7-22). Hover the mouse over each

coordinate to make it active (it will turn red), overtype a number, and then click anywhere on the workplane. The block will slide that distance (see Figure 7-23). If instead of (0,0) there are other numbers, the block isn't at the ruler's origin. You can overtype those numbers with zeros if you want to place it there.

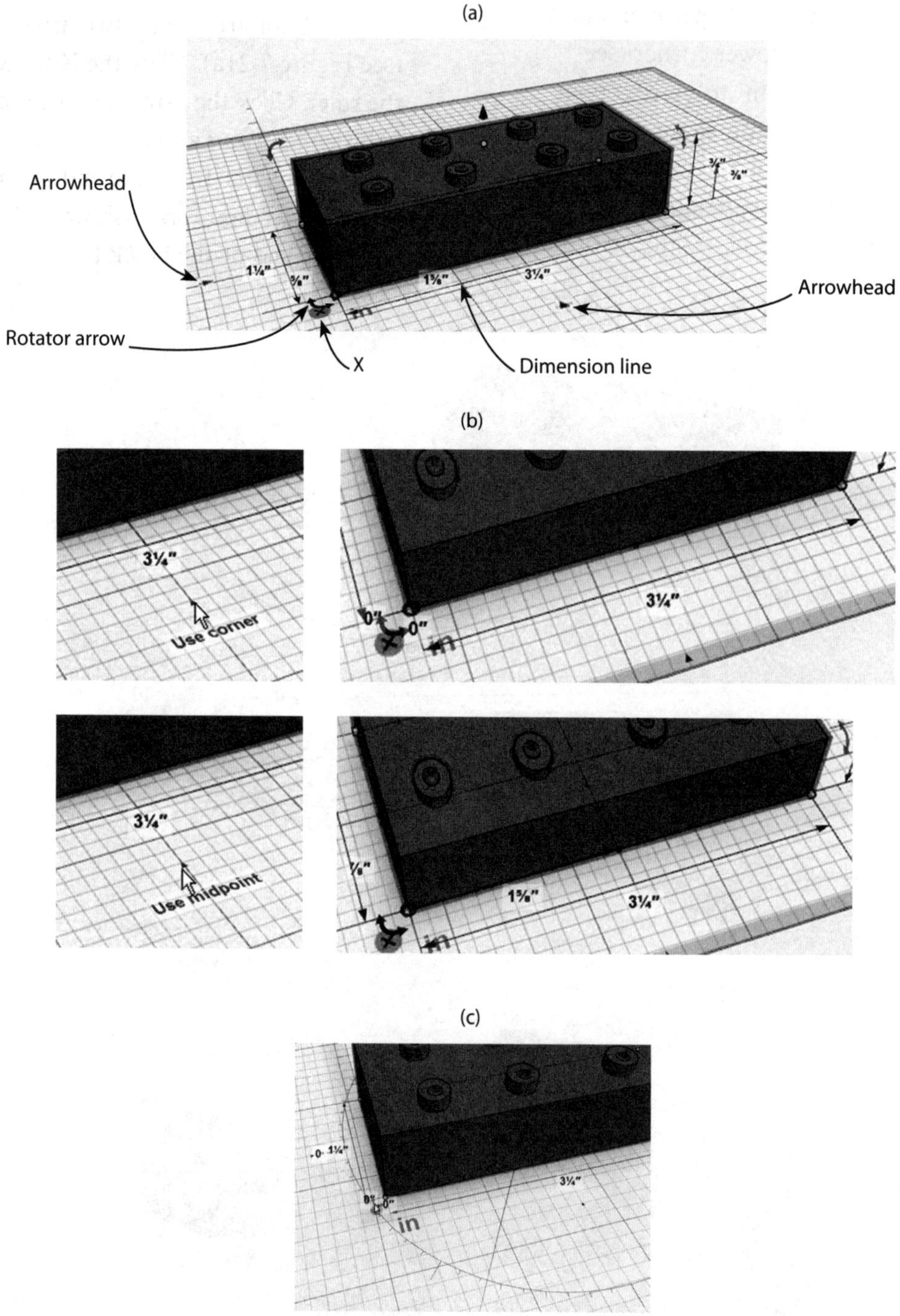

(a)

(b)

(c)

Figure 7-21 (a) Click the block to see its dimensions and manipulators. (b) Toggle between corner and midpoint to see different dimension placements. (c) Click the arrow to rotate the block.

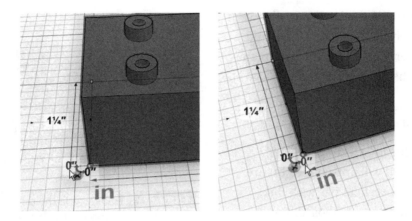

Figure 7-22 (0,0) is the ruler's origin. Hover over each zero to select it. Select and overtype it if you want to change the block's location on the workplane.

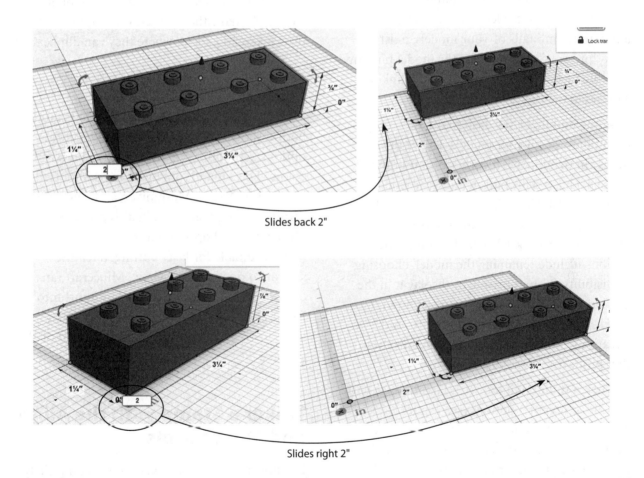

Slides back 2"

Slides right 2"

Figure 7-23 Overtype the 0, and the block will slide to the new coordinate location along the x or y axis.

Each dimension number—the distance between the endpoints' arrowheads—can be clicked and overtyped. After overtyping, click the workplane, and the block will change size (see Figure 7-24).

The Dashboard

When you exit the file by selecting Design | Close, your dashboard will appear (see Figure 7-25a). At the top is a vertical bar with links to editing your profile, the dashboard, the Tinkercad gallery (this is separate from the 123D gallery), video tutorials, a search field for the Tinkercad gallery, and a sign out link. On the left is a vertical bar that contains a search field for just your models (called "designs"), a link to thumbnails of your models, a list of your "projects" (the term for the model plus any file uploaded to accompany it), and a link to Tinkercad on Twitter.

In the middle of the dashboard are thumbnails of all your models. Hover the mouse in the upper-left corner of a thumbnail to see a Tinker This button and gear icon (see Figure 7-25b). Click the gear icon to edit the model's properties, duplicate it, move it to another project, or delete it. The Properties options include renaming the model, choosing its visibility (keep it private or display it in the Tinkercad gallery), and choosing a copyright license. The copyrights choices are from the Creative Commons, a nonprofit organization that promotes open education.

Click a thumbnail too see a larger image with download (export) options (see Figure 7-25c). The Tinker This button opens the file. Note the Creative Commons icons in the lower-right corner. Click them for information about this model's copyright choices.

To share a Tinkercad file with a friend, just copy and e-mail its URL. Multiple people can view a file simultaneously, but only one can work

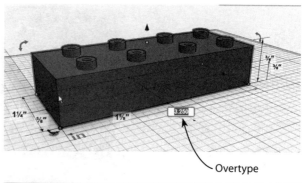

on it; a pop-up will appear that asks whether the viewer wants to reclaim rights to work on it. To let friends make the file their own, upload it to the Tinkercad gallery, where they can duplicate it.

Figure 7-24   Overtype the dimension numbers to change the block's size.

Download Options

When you click the Download for 3D Printing button shown in Figure 7-25c, you'll get four options: .stl, .obj, .svg, .x3d colors, and .xrml colors. The .stl and .obj files can be imported into 123D Design, Meshmixer, and Make for further development. You'll also get an .svg (2D) download option that provides a cross-section suitable for laser cutting, a feature similar to one in 123D Make. Minecraft fans can download a schematic file to import into MCEdit (the Minecraft editor). A link that has more instructions about that is at the end of this chapter.

Tinker a Stamp with the Align and Mirror Tools

For this project, we'll tinker an ink stamp that says "Be Happy." Set the units to inches and the snap to 1/16". This small snap allows finer movement than larger snaps do.

To make the stamp base, drag a cube from the Geometric library. Make it a 4"-long and

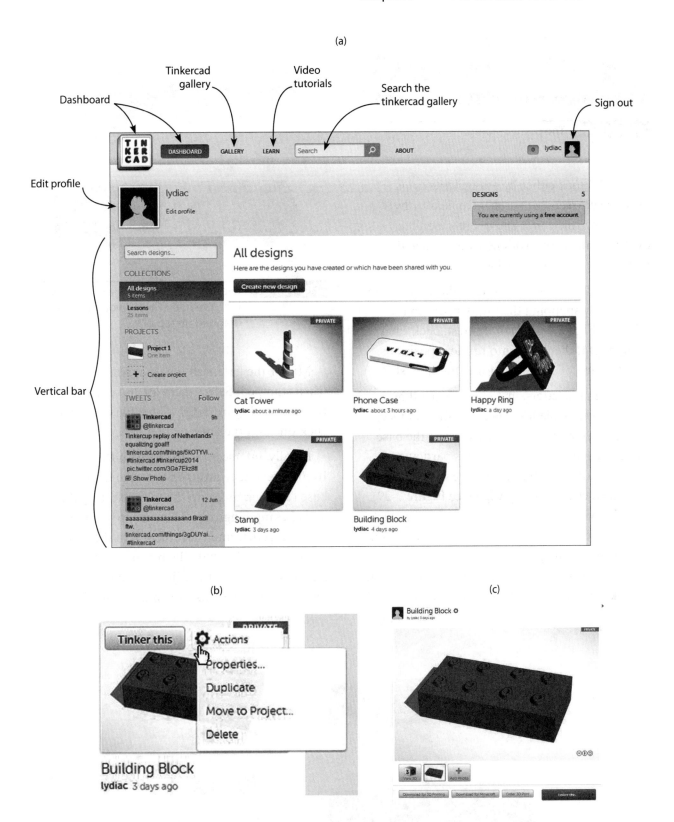

Figure 7-25 (a) The Dashboard. (b) Hover the mouse in a thumbnail's upper-left corner to see the Tinker This button and gear icon. (c) Click the thumbnail to see a larger image of the model and download options.

Figure 7-26   The printed stamp.

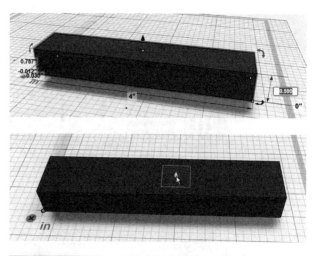

Figure 7-27   The cube is turned into a 4"×1/2" base with the ruler. Then a workplane is dragged on top of it.

1/2"-high block either by dragging its grips until the correct-size numbers appear or by putting the ruler on it and overtyping the existing dimensions with new ones. Then drag a workplane on top of the block (see Figure 7-27).

Drag letters from the Letters library and line them up by snapping their undersides to the grid (see Figure 7-28). For interest, click each to highlight, then drag into random positions. Rotate some of the letters for extra interest.

To align the letters, first select them all by clicking each while holding down the SHIFT key. Using a crossing window would select the base along with them, although in this particular case it wouldn't matter if the base got selected, too. Next, click Adjust | Align (see Figure 7-29). Handles will appear; hover the mouse over each for a preview of the alignment (see Figure 7-30). Then click on the one you want (see Figure 7-31). You might want to group the letters at this point to make operating on them together easier.

Mirror the letters so they'll appear properly when stamped. Select them all and click Adjust | Mirror (see Figure 7-32). Three handles appear; click one and the letters will mirror.

Finally, intersect the letters with the base. If you previously grouped both the letters and base together, ungroup them by selecting them and clicking Ungroup. Then group just the letters together and push them into the base (see Figure 7-33). Protrude them enough to make a clear stamp. Finally, adjust the letters' height relative to the base, as shown in Figure 7-33.

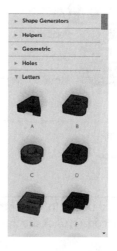

Figure 7-28   Drag letters onto the base, move and rotate them.

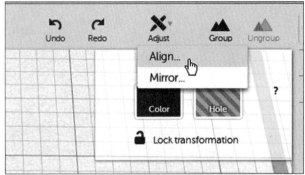

Figure 7-29 Select the letters and click Align.

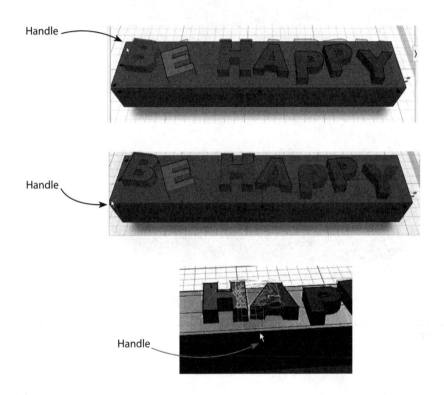

Handle

Handle

Handle

Figure 7-30 When Align is clicked, handles appear; hover over them for previews of the alignment.

Aligned

Figure 7-31 Click the alignment wanted.

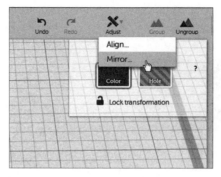

1. Click Mirror.

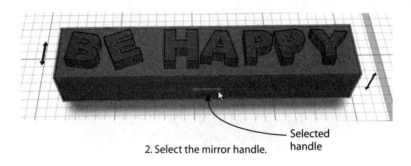

Selected
handle

2. Select the mirror handle.

3. Click the handle and the letters will mirror.

Figure 7-32 Mirroring the letters.

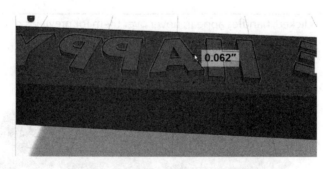

Figure 7-33 Push the grouped letters into the base to adjust their height.

The letters don't need to protrude much from the base, just enough to make a clear stamp. If you've previously grouped the letters with the base, ungroup them by selecting the group and clicking Ungroup. The letters need to be grouped with each other, however, and then pushed down into the base.

To be fully functional, the stamp needs to be printed with a rubber-like material to absorb ink or be paired with a self-setting rubber product such as Sugru. When using the latter, you would push non-mirrored letters into the block instead of protruding them from it, and mold the Sugru over it.

Tinker a Ring with an Imported SVG File

In this project we'll make the ring shown in Figure 7-34. To begin, set the grid properties to inches and set the snap to 1/16". Now follow these steps:

1. Drag a torus from the Geometric library onto the workplane and rotate it 90° (see Figure 7-35). See the plane that appears in the center of the ring? Drag a cube out and snap to it.

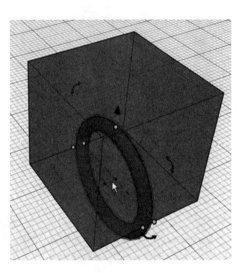

Figure 7-34   A ring made with an imported .svg file.

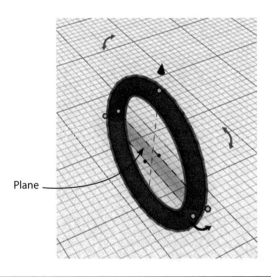

Plane

Figure 7-35   Drag a torus to the workplane and rotate it. Then snap a cube to the plane in its center.

2. Grab the small arrowhead at the top of the
   cube to move it to the top of the torus, and
   then drag down the white grip in the cube's
   middle to make it thinner (see Figure 7-36).
   Next, drag the white grip inward while
   holding the SHIFT key to make it smaller and
   maintain its proportions.

3. Make an .svg file like the one shown in
   Figure 7-37. This is easily done in Inkscape,
   a free illustration program. Download it at
   www.inkscape.org. I used the calligraphy
   tool to sketch "Be Happy" and then saved
   the file as an .svg. Tinkercad can only import
   small files, so keep your sketch simple and
   under 10KB.

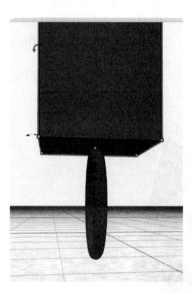
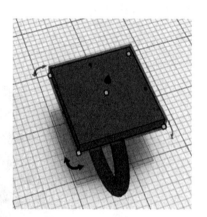

Figure 7-36   Move the cube to the top of the ring and then adjust its size.

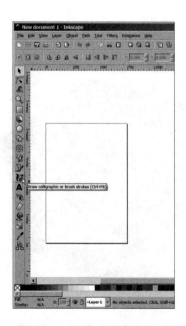
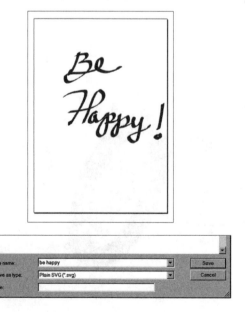

Figure 7-37   This sketch was made in Inkscape and saved as an .svg file.

4.
be happy.svg

On the menu panel, click File | Import File. Navigate to the .svg file (its icon will probably look like the Firefox or Chrome graphic) and then click Import (see Figure 7-38). The file may import very large, so zoom out if you can't see it properly. Note that the .svg file entered as an extrusion, and you can adjust it to any thickness desired.

5. Scale the .svg file by dragging the middle white grip inward while holding down the SHIFT key to maintain proportion. Drag a workplane onto the ring and then drag the .svg file onto it. You may need to fine-tune the scaling and placement (turn off the grid snap if needed). Then push the .svg file into the ring a bit with the arrowhead (see Figure 7-39).

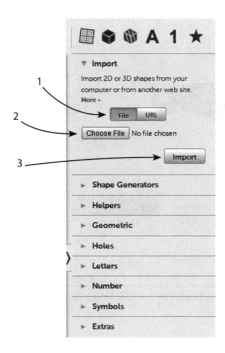

Figure 7-38   Import the .svg file via File | Import File on the menu panel. Here it has imported very large.

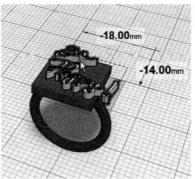

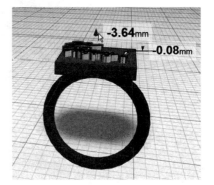

Figure 7-39   Scale the .svg file, move it to the ring, and then push it into the ring.

If you want to adjust the ring's size further, each part will have to be scaled separately, even when grouped. Remember to drag a workplane on the part before dragging the ruler out.

Tinker Other People's Models

Building on other people's models can save lots of time. Browse the Tinkercad gallery at www. tinkercad.com/things/. When you see something you like, click it and then click the Copy & Tinker button at the bottom (see Figure 7-40).

This opens the model in a new file that will be saved to your dashboard. Be aware that its original history doesn't copy with it.

Import an STL File of a Phone Case and Customize It

How does a phone case with your name sound? Browse the 123D Gallery (www.123dapp.com /search/gallery) or www.thingiverse.com (see Figure 7-41) to find a blank one. Download its .stl file and import it into Tinkercad the same

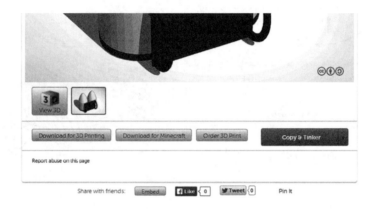

Figure 7-40 Find a model in the Tinkercad gallery, and then click the Copy & Tinker button to make it yours.

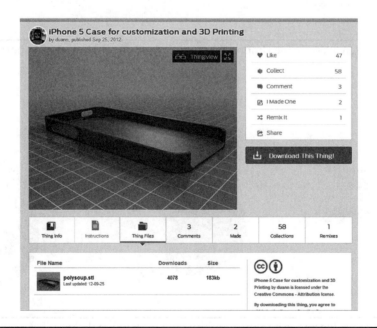

Figure 7-41 A phone case from Thingiverse.com.

way you imported the .svg file: click Choose File, navigate to it, and then click Import (see Figure 7-42). Now you can make it your own! I'll make this one my own by cutting my name into it. Here are the steps:

Flip the case, drag a workplane onto top of it, and drag letters onto the workplane from the Letters library (see Figure 7-43). Set a 1/8" grid snap to make aligning the bottom of the letters to the grid easier.

Move the letters into place, as shown in Figure 7-44. Select them by clicking each while holding down the SHIFT key. You can group them at this point if you'd like, which will make

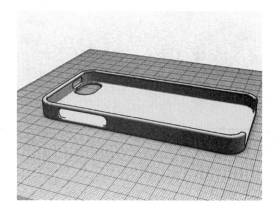

Figure 7-42   Import the .stl file into Tinkercad.

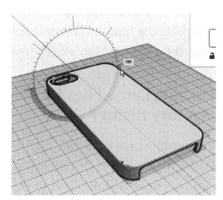

1. Flip the case.

2. Drag a workplane onto it.

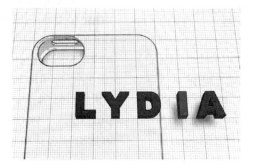

3. Drag letters onto the case.

Figure 7-43   Adding letters to the case.

Figure 7-44   Rotate the letters, move them into place, and align if necessary.

operating on them as a whole easier. Next, rotate and drag them to the desired location; align to finesse them if needed.

To cut the letters out, we first need to push them into the case (see Figure 7-45). They must intersect through the whole depth of the case, not just sit on top of it or pushed a little bit into it. Select all the letters (or the group of letters) and click Hole from the Inspector box. This turns the letters into letter hole parts. Group the hole parts with the case, and the hole parts turn into holes (see Figure 7-46).

Import an .Stl File Directly from the Web

You don't have to download an .stl file before importing it into Tinkercad. You can import it with just an URL address (see Figure 7-47)! Find something you want from Thingiverse and open its page. Click on the Thing Files folder at the bottom of the page to make the .stl link appear. Then right-click on the .stl link and click Copy Link Address.

Back in Tinkercad, click on Import | URL and paste the link into the text field. Click the Import button and the .stl file will appear.

Manage Large Files by Removing Memory History

Tinkercad works best with small files. When a file becomes large and complex, it runs slowly. One way to deal with this is to erase the memory history. This can be done by duplicating the file because duplicate files don't retain any memory prior to the duplication. Another way is to download the file, or a complicated part of it, as an .stl and reimport that .stl into the file or into a new file. All memory prior to the reimport will be lost, making the .stl, for all practical purposes, a primitive geometric form. Both these actions are done through the dashboard.

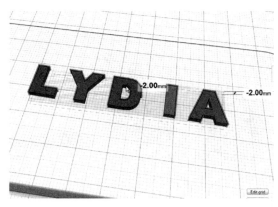

1. Intersect the letters with the case.

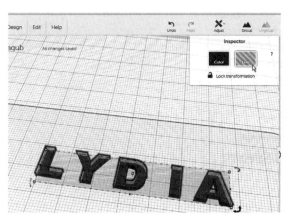

2. Select them and turn them into hole parts.

3. Group the hole parts and the case together.

Figure 7-45 Cutting the letters out.

Figure 7-46 The customized phone case.

Thing Files folder —

.stl link

1. Click on the Thing Files folder
to make the .stl link appear.

2. Click on Copy Link Address.

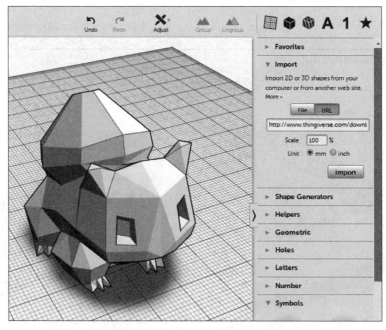

3. Paste the link address into File | URL.

Figure 7-47   Importing an .stl file via its URL.

Tinkercad as an STL File Repairer

Mesh models often have holes or non-manifold edges (an edge shared by more than two polygons). This makes them non-printable. The 123D Meshmixer app fixes such problems, so all the apps except Tinkercad are sent to it for printing preparation. Tinkercad does not send its files there because it heals bad .stl files itself upon import. It will import any .stl file no matter how bad its mesh is, and turn it into a watertight one. A quick way to fix unprintable .stl files is to import them into Tinkercad and export them out again. Note, however, that the model might get changed a bit in the process. If you don't like the result, or want more control over that result, import the .stl into Meshmixer and make the changes yourself. Know that Tinkercad does have a limit on polygon size. If your .stl doesn't import, it's probably too big.

Shape Generators

Shape Generators are parts that contain JavaScript programming. This enables you to edit them way beyond Tinkercad's simple tools. Shape Generators are dragged out of the Library panel like any other part, and can intersect, be grouped with other parts, turned into holes, and anything else the native parts can do. They're made by both the Tinkercad team and the user community. Although the Library panel has a nice selection (see Figure 7-48), you can find more in the Thingiverse (be sure to include "shape generator" in the search terms).

You don't have to know JavaScript yourself to edit a Shape Generator. The programmer defines how it can be manipulated and provides a control panel for doing so. Figures 7-49 through 7-52 show four edited Shape Generators.

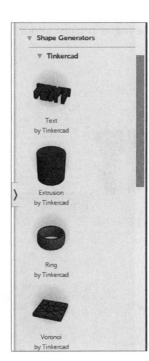
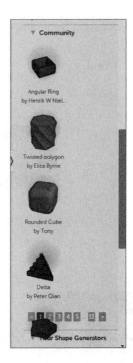

Figure 7-48 Shape Generator collections by the Tinkercad team and user community.

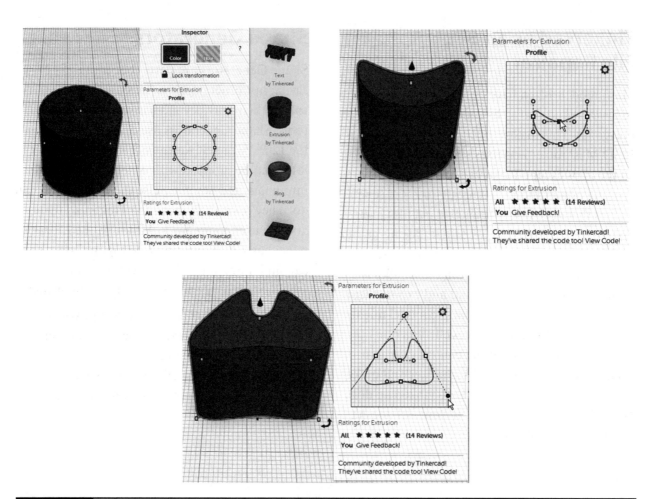

Figure 7-49 The Extrusion Generator can be made into just about any shape.

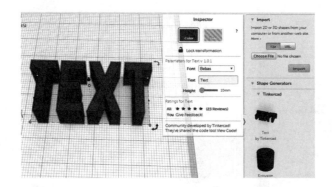

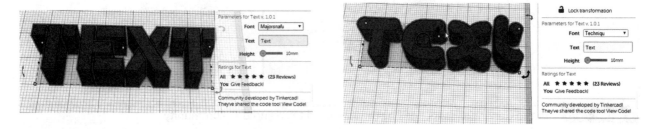

Figure 7-50 The Text Generator offers font choices.

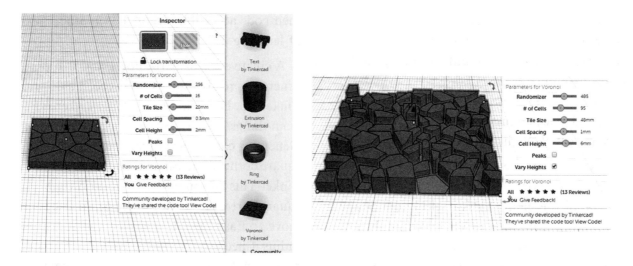

Figure 7-51   The Voronoi Generator creates organic shapes.

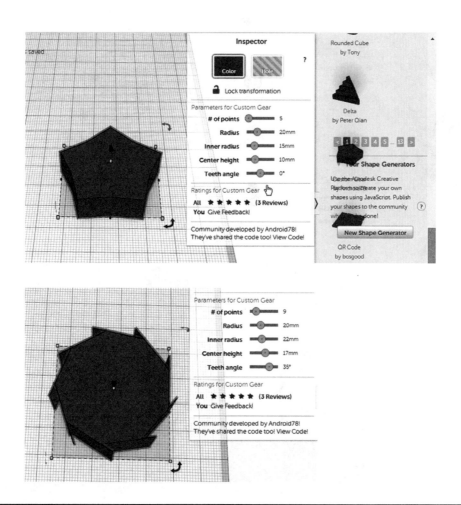

Figure 7-52   The Custom Gear Generator makes gears of all sorts.

Figure 7-53 shows a helix that's in the community collection. I adjusted it with the control panel and made it taller by dragging the arrow manipulator. Then I dragged a cylinder out from the Geometric library, scaled it tall and thin, and placed it inside the helix for a perch that any cat would like to hang out on.

If you're interested in making your own Shape Generators, a link with information is at the end of this chapter. Also, select Your Shape Generators | New Shape Generator and then click one of the shapes to see its code, which you are encouraged to edit (see Figure 7-54). Many Shape Generator programmers make their code accessible. Study it as another means of learning it.

A creative use of parts, holes, and shape generators really extends what you can do with Tinkercad. For example, at https://tinkercad .com/things/la63cSw4tbV is a user-made Shape Generator called Easy Fillets that rounds off square corners (see Figure 7-55). Click the Tinker This button to make it your own. Make a square part with the Rectangle geometric shape, move an Easy Fillet to a corner, and adjust the radius if needed. Group the Easy Fillet and the part, click the workspace to exit, and the part now has a round corner (see Figure 7-56).

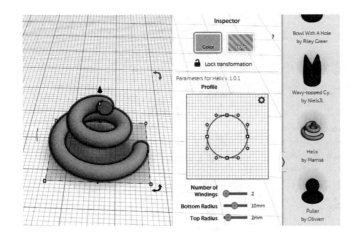
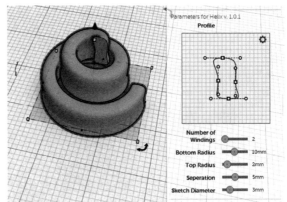
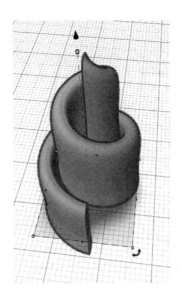
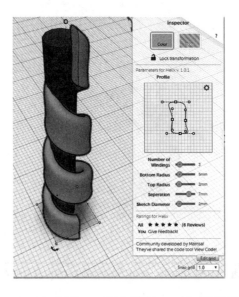

Figure 7-53   A cat perch made with a Shape Generator helix.

Figure 7-54   The Your Shape Generators | New Shape Generator collection makes each shape's code accessible for study and editing.

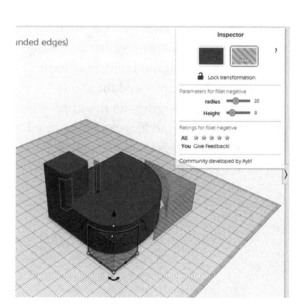

Figure 7-55   A user-made Shape Generator called Easy Fillets.

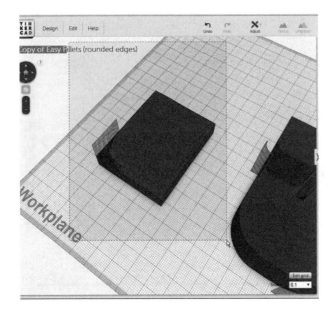 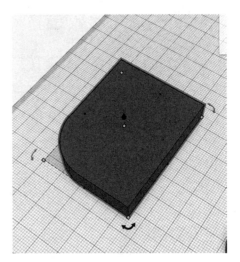

Figure 7-56 Using the Shape Generator to make a square corner round.

Summary

Tinkercad is a simple, powerful, web app designed especially for those with no solid modeling experience. Models, called "designs," are made by pushing shapes together and using the hole, align, grouping, and mirroring tools on them. Shape Generators and the ability to import .stl files greatly expand this app's capabilities. All Tinkercad files can be exported as .stl files, which enables them to be further developed in other programs.

Sites to Check Out

- **Tinkercad help forum** http://forum.123dapp .com/123d/products/123d_tinkercad

- **Tinkercad blog** http://blog.tinkercad.com/

- **Twitter handle** @tinkercad

- **Keyboard shortcuts** http://blog.tinkercad. com/keyboard-shortcuts/

- **Information about Creative Commons and its licenses** www.creativecommons.org

- **Information on Shape Generators** http:// blog.123dapp.com/2013/12/sketch-and-text

- **Information on exporting Tinkercad files into Minecraft** http://blog.tinkercad .com/?s=export

- **Shape Generator information for developers** http://api.tinkercad.com/ libraries/1vxKXGNaLtr/0/docs/index.html

Cut It! with 123D Make and the CNC Utility

SO YOU WANT TO MAKE a 3D item by assembling cut pieces of cardboard, acrylic, wood, or sheet metal? Then 123D Make is for you. Although Make files can be sent to a 3D printer, this app is primarily used to slice .stl and .obj models into manufacturable cutting patterns that can be read by computer-operated machines. It also generates a file that can be read by the Cricut Explore, an electronic device that cuts designs out of paper, card stock, poster board, vinyl, and fabric.

Cut patterns have diverse uses; for instance, they can be a craft kit, terrain model, piece of sculpture, or furniture prototype. Figure 8-1 shows examples from the 123D Gallery. The chairs are made from glued-together stacks, and the Snuffleupagus is made from folded panels;

the bull's stacked sheets include two dowels. Make even generates assembly instructions showing how the cut pieces go together.

Perhaps you just want to carve your ready-made pattern out of wood, and not by hand. Then the CNC Utility is for you. It turns a design into a file that a CNC (computer numeric controlled) machine can read. That's a machine operated by a personal computer that holds the cutting tools as well as moves, cuts, drills, and shapes wood. It could be the same computer you did your drawings on.

In this chapter we'll look at both apps, starting with Make. At www.123dapp.com/make you'll find three platforms: iPhone/iPad, Web, and PC/Mac desktop. Download the desktop app because it has the most features.

Dowels

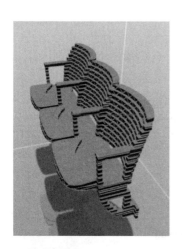

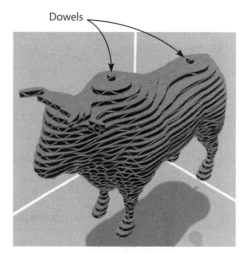

Figure 8-1 Examples of .stl files turned into sliced assemblies with 123D Make

Getting a Make-Importable File

First, you need a Make-importable file. 123D Make imports .stl and .obj files. They should be finished products because there are no tools for editing models in Make. In fact, if your .stl or .obj file won't import into Make, it may be due to holes, manifold edges (edges shared by more than two polygons), or some other flaw that makes it nonprintable. In this case, import it into

Meshmixer for analysis and repair (discussed in Chapter 9). Here's how to get a Make-importable file from the different 123D apps (see Figure 8-2):

- **123D Design** You can send a .123dx file directly from Design to Make. You can also select just a portion of the Design file and send it to Make through the glyph that appears when a model is selected. This enables different fabrication options on the model, such as stacked slices on one part and interlocked

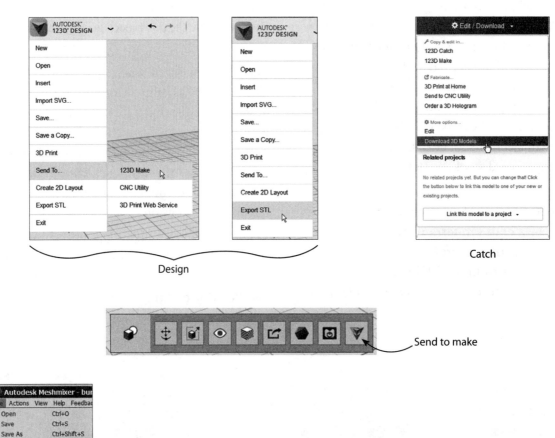

Design

Catch

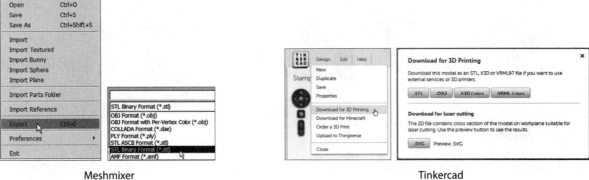

Send to make

Meshmixer

Tinkercad

Figure 8-2 How to turn 123D app files into Make-importable files

slices on another. Alternatively, export the .123dx file to an .stl file to your computer.

- **123D Catch**   The .3dp file is in your online 123D account. Click Models, select the specific model, and then click the Edit | Download button to download .stl and .obj files to your computer.

- **123D Meshmixer**   Export the .mix file to your computer as a binary (not ASCII) .stl file.

- **Tinkercad**   Download the .stl file to your computer.

The 123D Make Interface

Once you launch Make, you'll see the interface shown in Figure 8-3. It consists of a workspace, menu bar at the top, and vertical panel on the left. The menu bar has a drop-down arrow that accesses a submenu (see Figure 8-4). It also has Undo and Redo arrows as well as links to your online account/resources (see Figure 8-5). There are also icons to rotate the model and see the changes made to it. Those are grayed out because we haven't imported a model yet.

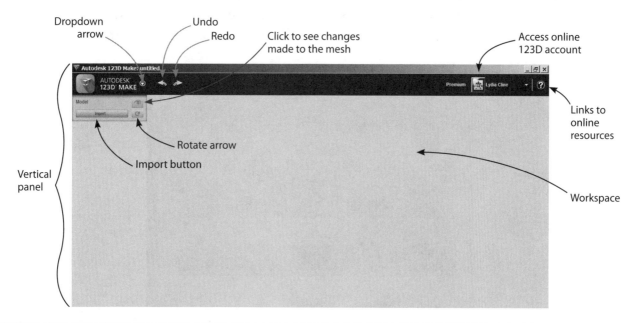

Figure 8-3   The Make interface

Figure 8-4   Click the drop-down arrow in the menu bar to see this submenu.

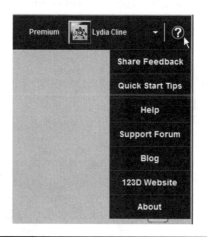

Figure 8-5   Links to your online account/resources are on the menu bar.

The menu bar's submenu contains these functions:

- **New**   This item closes the current file and makes the Import button appear.

- **Open Example Shapes**   This item brings up choices of premade files to experiment with.

- **Open**   This item brings up a browser window where you can open a project from your computer, online 123D account, or the 123D Gallery (see Figure 8-6). You need to be logged in to access it.

- **Save**   Click here to save a .3dmk (Make) file to your computer or to your online 123D account.

- **Save a Copy**   Click here to save a copy of the .3dmk file to your online 123D account. It will become the current file.

- **Export Mesh**   This item turns the .3dmk file into an .stl or .obj file.

- **3D Print**   This item sends the .3dmk file to Meshmixer for printing preparation.

- **Exit**   This item closes the file and exits the app.

Click the Import button on the vertical panel to bring up a navigation browser (see Figure 8-7). We'll navigate to and import the glass head capture made in Chapter 5 (see Figure 8-8). If the model enters at an odd angle, as in the top

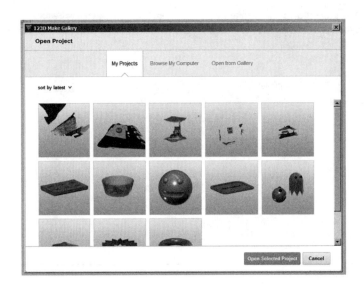

Figure 8-6   Click Open to navigate to a file on your computer, online account, or 123D Gallery.

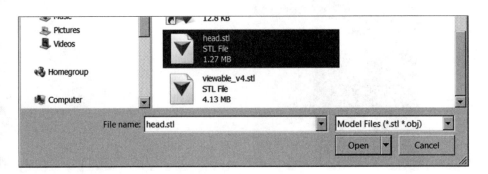

Figure 8-7   Click Import and browse to an .stl or .obj file.

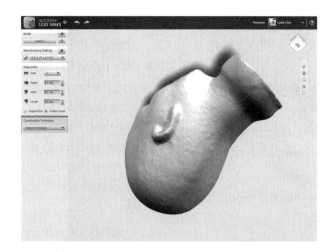

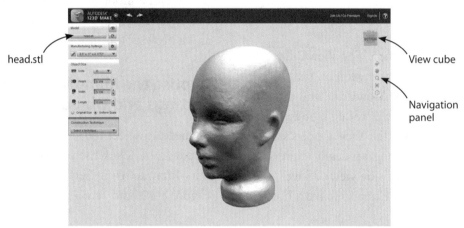

head.stl

View cube

Navigation panel

Figure 8-8   An imported .stl file of a glass head. It imported at an odd angle and was straightened with navigation tools.

graphic, it's because the program it was imported from doesn't share the same coordinate system as Make. Adjust it by clicking the rotate arrow in the vertical panel or by using navigation tools. To that end, note the ViewCube and navigation bar that now appear in the workspace (see Figure 8-9).

Navigate the Interface

You can move around the interface in one of three ways:

- **By using the navigation bar**   The navigation bar contains tools for tumbling, panning, zooming, and framing the model. They are:

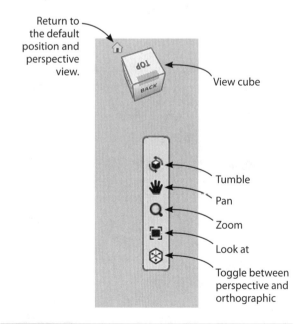

Return to the default position and perspective view.

View cube

Tumble

Pan

Zoom

Look at

Toggle between perspective and orthographic

Figure 8-9   A ViewCube and navigation bar appear after a model is imported.

- *Tumble* Click and drag to move the model at any angle around the workspace.

- *Pan* Click and drag to slide the model around the workspace.

- *Zoom* Click and drag the cursor up and down for a view that's closer to or farther from the model.

- *Look At* Click to center the model and fill up the screen.

- *Toggle* Click to switch between perspective (3D) and orthographic (2D) views.

- **By using the ViewCube** The ViewCube shows the model's orientation on the workspace. Click the ViewCube and drag to rotate it. The model will rotate along with the ViewCube. Click the View Cube's sides to see the model orthographically (that is, as a 2D top, front, or side view). Hover the mouse over the ViewCube to make a house

icon appear. Click that icon to return the model to a perspective view.

- **By using the mouse** Right-click anywhere on the screen and drag the cursor to tumble the model (that is, move it at any angle around the workspace). Press down the scroll wheel to drag the cursor, which pans (slides) the model around the workspace. Roll the scroll wheel up and down to zoom in and out.

Options for settings also appear in the vertical panel after a model is imported. Note the button that says "head.stl." Clicking it makes the file close and a browser for importing a new file appear.

Choose the Manufacturing Settings and Object Size

We need to select the sheet size the project will be cut from. Click the pencil button to access the Manufacturing Settings box shown in Figure 8-10, and scroll through the choices. If

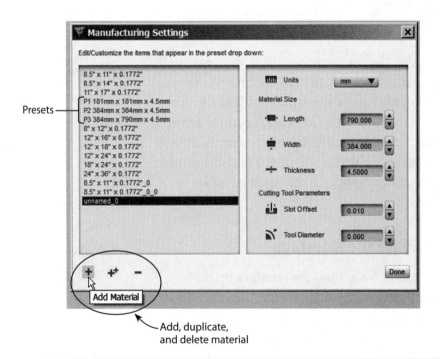

Figure 8-10 Click the pencil button to access the Manufacturing Settings box. Then, choose a sheet dimensions preset or create your own.

you'll be using a laser-cutting service through 123dapp.com, choose preset P1, P2, or P3. To make a variation of a preset, click that preset to highlight it and then click the double-plus sign icon to duplicate it. Enter the new settings. To delete a preset, highlight it and click the minus sign icon.

This screen also has graphics at the bottom for adding your own presets, which is handy for nonstandard materials. Click the plus sign icon to add a new preset, then enter the units, slot offset, and length, width, and thickness of the material into the text fields.

Now select the model's physical size. Choose its units and adjust the height, width, and length if needed. Clicking the Uniform Scale button scales the file in all directions. When it's unchecked, you can scale the model differently along the three axes. Clicking the Original Size button reverts the file back to the size at import. Obviously, the larger the file is, the more slices that will be generated, and the more sheets you'll need.

Construction Technique

Now the fun starts! At the bottom of the panel is the Construction Technique box. There you'll find six types: Stacked Slices, Interlocked Slices, Curve, Radial Slices, Folded Panels, and 3D Slices.

Each technique has at least one option unique to it, but most options overlap between them. Click the drop-down arrow to select a technique (see Figure 8-11). The model will update to reflect it, with accompanying 2D graphics of slices laid out on sheets in the Cut Layer tab to the right. As you finesse the options, the slices automatically update. Be aware that *slices* and *parts* are not synonymous. A slice can consist of multiple parts.

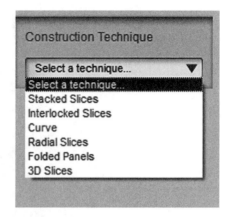

Figure 8-11 Click the drop-down arrow in the Construction Technique box to see the choices.

Use Stacked Slices for the Make Process

Let's go through the whole Make process, from choosing the technique to printing the file. We'll use the Stacked Slices technique. Here are the steps to follow:

1. Click the Stacked Slices option technique. This makes cross-sectional slices for gluing and stacking on top of each other (see Figure 8-12). The Dowels option creates pegs that help align and hold the slices together, and it lets you choose their size, location, and shape (see Figure 8-13). Move a dowel by highlighting and dragging it; delete it by highlighting and pressing the DELETE key. Change the slice direction, as shown in Figure 8-14, by dragging the blue handle. The handle can only be dragged when it's visible (dark blue), not hidden (muted) behind the model. Click the Model Issues button to see problems such as unconnected parts; they'll show up in blue (see Figure 8-15).

    Sometimes problems can only be fixed by changing the model somewhat. In that case, click the Modify Form icon. Three buttons will appear at the bottom of the screen:

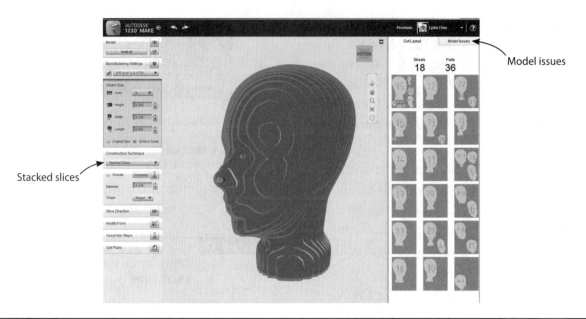

Stacked slices

Model issues

Figure 8-12 The Stacked Slices option turns the form into cross-sectional slices.

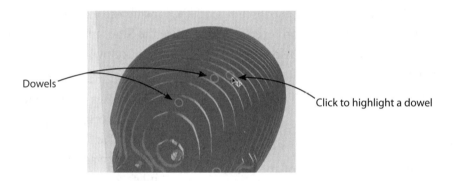

Dowels

Click to highlight a dowel

Figure 8-13 Adjust dowel locations by dragging them.

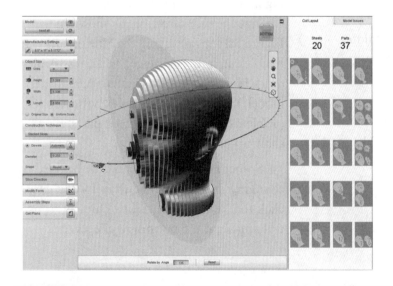
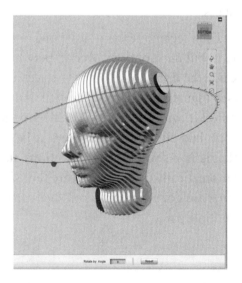

Figure 8-14 Change the slice direction by dragging the blue handle.

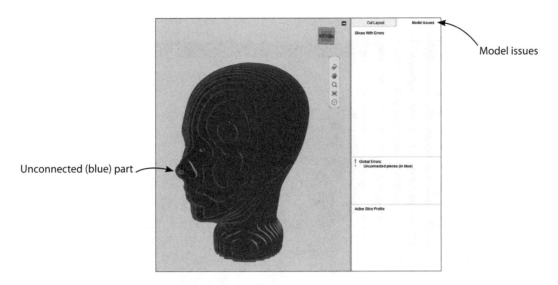

Unconnected (blue) part

Figure 8-15 Click the Model Issues button to see problems. Here, the blue part is unconnected to the rest of the slice.

Hollow, Thicken, and Shrinkwrap (see Figure 8-16). Hollow shells out the model, reducing the amount of material needed to build it. Thicken widens cutouts that are too thin to fabricate. Shrinkwrap approximates and smoothes details that are too fine to cut out. It also closes holes, which is particularly useful on models made with 123D Catch. Select one, adjust its slider, and click Done.

Be aware that Shrinkwrap affects the whole model, not just a problem slice. Therefore, the end product may not look exactly like the model. Additionally, you should know the machine's limitations; for instance, a router can't make square corners. Appropriate clearances are needed, which is the space for parts to connect.

2. Click the Assembly Steps icon and choose the material (see Figure 8-17a). Drag the mouse along the slider at the bottom of the screen to watch an animation putting the parts together. To view an assembly sheet, click its thumbnail in the Assembly Reference panel (see Figure 8-17b). Move the sheet by clicking and dragging; scroll the

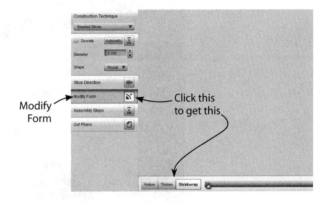

Modify Form — Click this to get this

Figure 8-16 The Modify Form options may be able to fix problems inherent in the model.

mouse wheel to zoom in on a part; click the X in the upper-right corner to return to the model.

Study the graphics in the Cut Layout tab. They show how many sheets of material are needed, how many slices and parts there are, and where problems lie. Hyphenated labels, such as Z-3-2, describe the part's axis (Z), the slice number (3), and part number (2). Blue outlines are the model's outside edge;

(a)

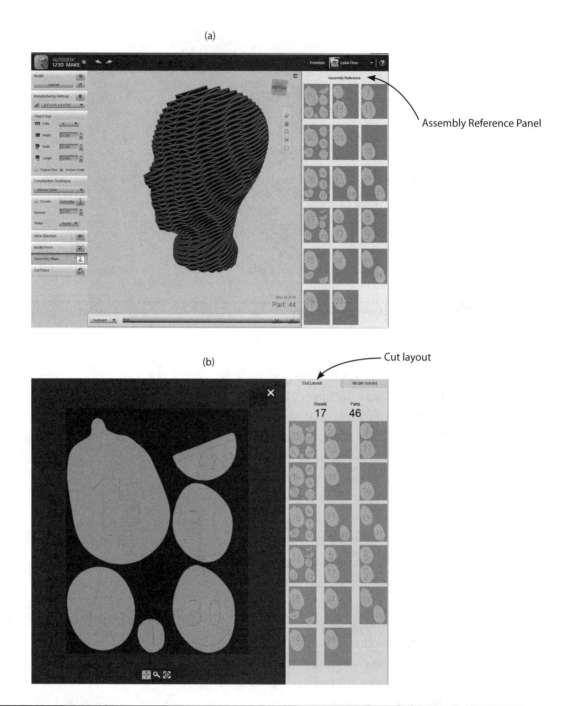

Assembly Reference Panel

(b)

Cut layout

Figure 8-17   (a) Click the Assembly Steps icon to choose the material. (b) See an animation of how to put the parts together.

green outlines are cuts inside the model for hollows; yellow outlines are scored guides for placing parts during assembly.

The parts are arranged to use as much of each sheet as possible, and all are oriented the same direction. Currently, there is no other way to arrange them. If you want to change the arrangement—perhaps the wood grain faces the wrong direction—import the file into another vector program such as Inkscape or Illustrator and change it there.

3. Export the file by clicking the Get Plans icon, choosing the format (.eps, .pdf, or .dxf) and units, and clicking Export (see Figure 8-18). The file will be exported into a format ready to import into a CNC machine or Cricut. The .eps option creates a ZIP folder with separate files for each sheet. This option, as well as the .dfx option, puts text on one layer and profiles on another, which is useful when laser cutting. The .pdf option creates one file that puts all the slices on separate pages (see Figure 8-19). Its advantage is that it can be

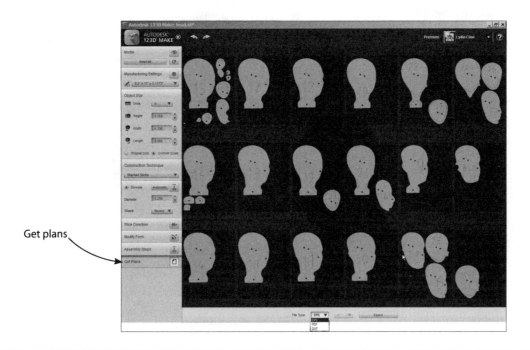

Figure 8-18 Click the Get Plans icon, choose a file format, and export it.

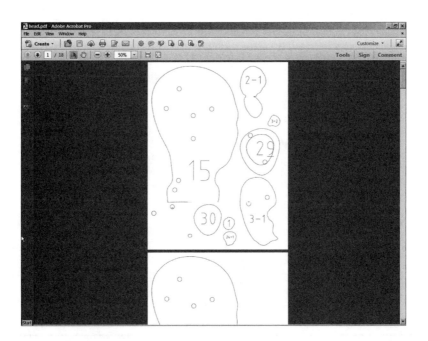

Figure 8-19 An exported .pdf file of the slices

viewed with the ubiquitous Adobe Reader, whereas you may not have an program that can read .eps files.

Interlocked Slices

This technique slices the model into two stacks of slotted parts that lock together in a grid. It uses less material than the Stacked Slices technique. Problem slices show up in red (see

Figure 8-20) and may be fixed by adjusting the options shown in Figure 8-21. Those options include letting you adjust the slice distribution, notches, and relief. You can distribute the slices by count, distance, or a custom setting. The 1st Axis field sets the number of slices in each direction, and the 2nd Axis fields set the distance between slices. Increase the numbers to give the model a more defined form or decrease them to use less material. Click the horizontal bars

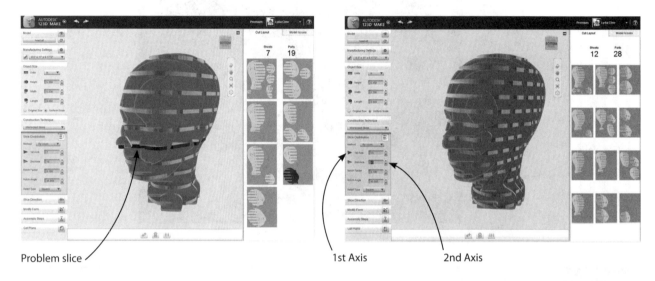

Problem slice

1st Axis   2nd Axis

Figure 8-20 The Interlocked Slices technique. A problem slice is indicated in red and may be fixed by adding slices via the 1st Axis field.

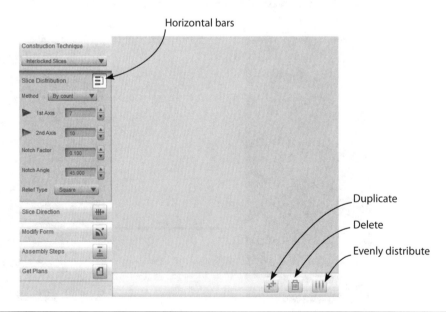

Horizontal bars

Duplicate

Delete

Evenly distribute

Figure 8-21 Options for the Interlocked technique. Click the bars to bring up the Duplicate, Delete, and Evenly Distribute buttons.

to bring up graphics that duplicate, delete, and evenly distribute the slices. Click any slice and drag it to change its position. Click the Evenly Distribute button after moving individual slices to evenly space them. To add a slice, highlight an existing one and click. A duplicate will appear next to it that follows the model's contours. To remove a slice, highlight it and click Delete.

Large parts can twist, making assembly difficult or impossible. Make's default square-cornered slots may also cause assembly problems, depending on the design. Use the Notch Factor field to flair the slot by a specific amount relative to its width; use Notch Angle to set the angle relative to the slot's direction (see Figure 8-22). A 45° angle often works. Ensure that notches of any time are wide enough to be cut out.

The term *relief* refers to the notch shape. The default relief type is square, but you can choose a relief type of horizontal, vertical, or dog bone, which is a shape with no inside rounded corners. You can further edit the relief by specifying the tool diameter. Go back to the Manufacturing Settings, make a new preset, and enter a number

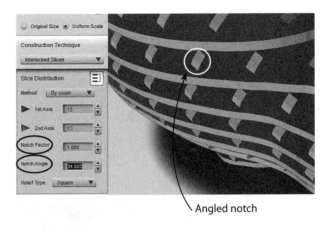

Angled notch

Figure 8-22 Notches are square by default but can be angled.

in the Tool Diameter field. Dog bones can't be generated when the Tool Diameter is set to zero.

Curve

This technique cuts slices perpendicular to a curve (see Figure 8-23). All the curves are on one plane. It works best on flowing, organic shapes, such as animals. Options are similar to those of

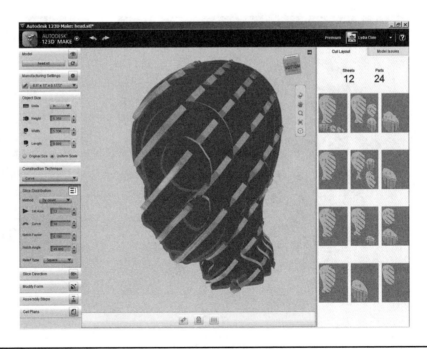

Figure 8-23   The Curve technique cuts slices perpendicular to a curve.

the Interlocked Slices technique. Additionally, you can bend the curve.

Radial Slices

This technique creates radiating slices from a central point. It's best used on round, symmetrical objects. As you can see from its effect in Figure 8-24, not all designs are suited for all techniques. Here, the default generation (top) returned multiple problem slices. Adjusting the count, 1st axis, and slice direction fixed them (bottom left), but turning the model revealed a new, unconnected part (bottom right).

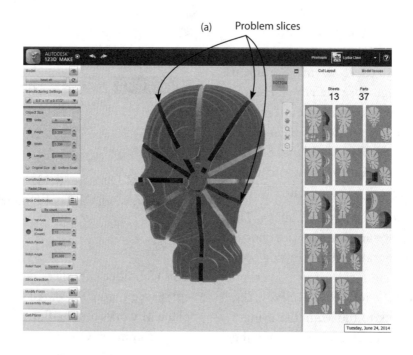

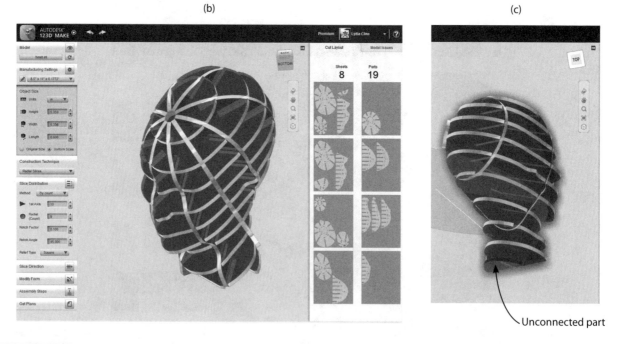

Figure 8-24 Slices in the Radial Slices technique emanate from a central point. This model displays multiple problem areas.

Folded Panels

This technique turns the model into 2D segments, or panels, of triangular meshes that are folded multiple times. Paper, cardboard, and sheet metal can be used with this technique.

Figure 8-25 (top) shows the default generated head. It has a lot of problem areas and a message that the parts are too large for the sheets. Clicking the Split Panels option turns every face into a panel, which fixes this but generates hundreds of parts. You can lower the

vertex and face count to simplify the model (a process called "decimation") to make assembly easier. Of course, this simplifies the model, too, as seen in the bottom graphic.. If you want to use the decimated model but preserve a copy of the file with its mesh intact, export it through the Export Mesh option in the menu bar (see Figure 8-26).

Click the Add/Remove Seams icon and then click a panel edge to add fold lines that divide large panels into smaller ones or remove fold

Add/remove seams

An error message warns that the generated parts are too large for the material.

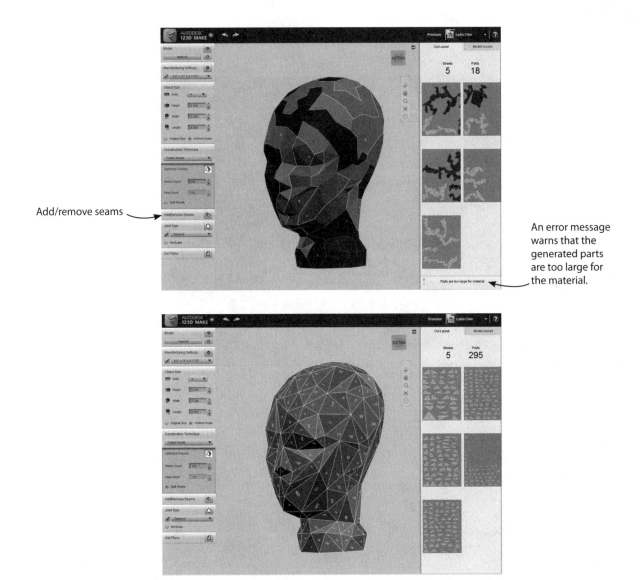

Figure 8-25 The top graphic shows multiple problem areas. The bottom graphic shows the model with split panels and lower vertex and face counts, which fixes the problems.

Figure 8-26 The menu bar's Export Mesh option preserves the mesh.

lines to combine small panels into bigger ones (see Figure 8-27). You can also turn fold lines into perforation lines. The perforations will be the thickness of the chosen material. Split Panels will deselect when you do this; click it again when finished if your model becomes covered with problem areas.

Joints for Folded Panels   Joints are tabs (which Make calls "ticks") that connect the panels. There are 10 types, each shaped differently. All have finessing options (see Figure 8-28). If a model has lots of problem areas, changing the joint type often helps.

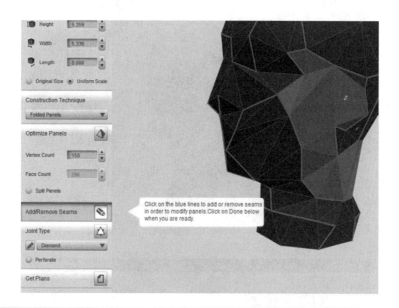

Figure 8-27   Click the Add/Remove Seams icon to make bigger or smaller panels.

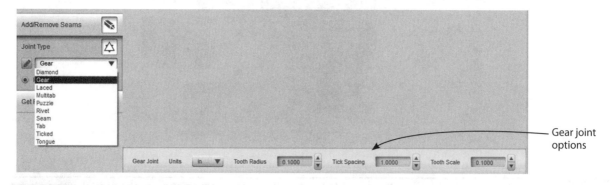

Figure 8-28   There are 10 joint types; the Gear joint and its options are displayed here.

Joint types are illustrated in Figure 8-29 and detailed in the following list:

1. **Diamond**   Fold and glue/weld the triangular tabs.

2. **Gear**   Fold and glue/weld the rectangular tabs (cut the dark area or areas out).

3. **Laced**   Entwine the sheets together.

4. **Multi-tab**   Fix the tabs together.

5. **Puzzle**   Fit pieces together.

6. **Rivet**   Pin tabs to one another.

7. **Seam**   Sew tabs together (like a sewing pattern).

8. **Tab**   Insert tabs into slots.

9. **Ticked**   Connect tabs along the seams.

10. **Tongue**   Insert a tab into slots.

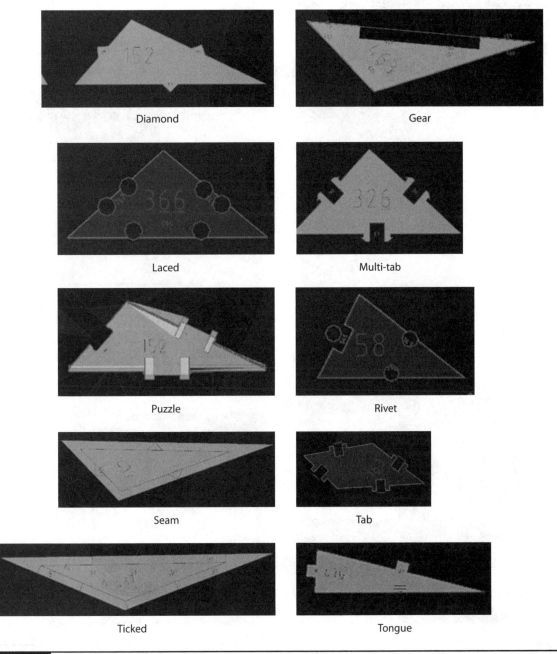

Diamond	Gear
Laced	Multi-tab
Puzzle	Rivet
Seam	Tab
Ticked	Tongue

Figure 8-29   Folded panels have 10 joint options.

3D Slices

This technique slices the model. However, unlike Stacked Slices which merely step to create a form, 3D slices actually follow the form. That is, slices are made along all the x, y, and z axes (all three of the model's physical dimensions). See Figure 8-30.

Save and Export the Model

You can save your work on your computer or to your online 123D account. When saving it online, as shown in Figure 8-31, you'll be asked to describe it. You also choose between keeping it private and making it public, which displays it in the 123D Gallery. You can always edit these

Figure 8-30 The 3D technique makes slices that follow the model's form.

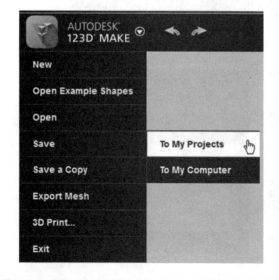

Figure 8-31 Click the drop-down arrow on the menu bar to save the file. Saving to My Projects brings up a dialog box asking for more information.

choices later. You can also export the model to your desktop as an .stl or .obj file (refer back to Figure 8-26).

Send the Model to a Service Bureau

At the www.123dapp.com site, click 3D Printing and then More Making Options (see Figure 8-32). Clicking Ponoko's icon takes you to their site. Click Make at the top, click Make again, and then click Get Making (see Figure 8-33). After entering the project's details, you'll be quoted a price. The site has a tutorial video and a lot of information about their services. Know that the parts and slices will be mailed to you unassembled.

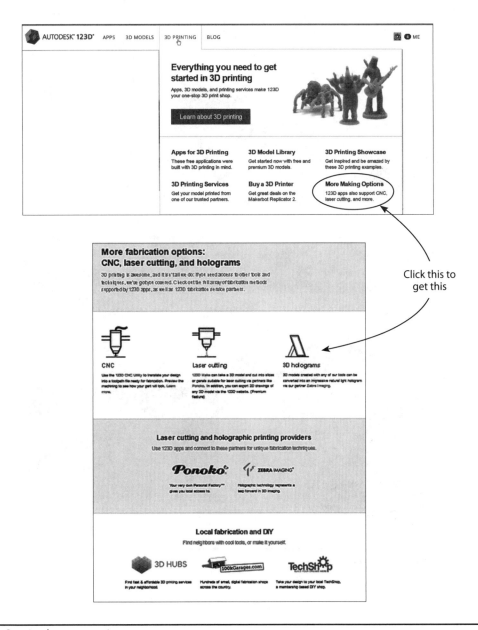

Figure 8-32 Service bureau options

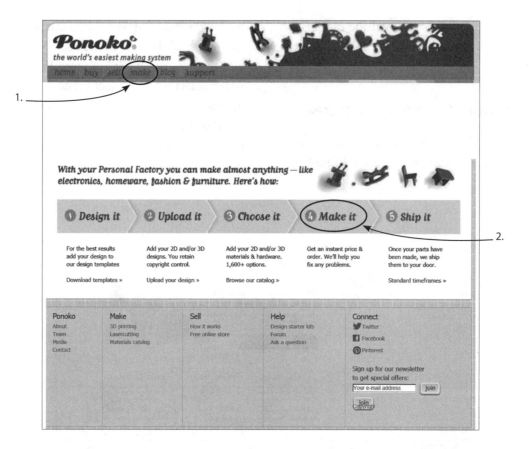

Figure 8-33   The order process at Ponoko

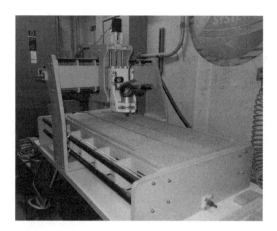

Figure 8-34 A CNC router is a tool that cuts sheets of wood, plastic, wax and foam. This router is in Hammerspace, a makerspace in Kansas City, MO.

Figure 8-35 The CNC Utility can be launched from inside 123D Design.

The CNC Utility

The CNC Utility is a web-based app for creating files that drive a CNC router. This is a tool that cuts sheets of wood, plastic, wax, cardboard, and foam (see Figure 8-34). The end result is a file of *G-code*, a language that computerized tools understand. G-code tells the tool where to move, how fast, and what path to take. You can then send this file to your personal router if you have one, or to an online service bureau or local hackerspace if you don't.

There are built-in settings for the whole ShopBot line, a company whose CNC lathes and routers are standard equipment in woodworking shops. However, all CNC routers are supported, so you can make a setting for a different router.

Launching the CNC Utility

The CNC Utility can be launched from inside 123D Design (see Figure 8-35) or opened directly to import an .stl, .obj, .svg, or .eps file. Point the Chrome browser to http://apps.123dapp.com/cnc/ and click Start a CNC Project (see Figure 8-36). A browser will appear with which to navigate to a file in your online account, on your computer desktop, in the 123D Gallery, or to a premade

Figure 8-36 At http://apps.123dapp.com/cnc/, click Start a CNC Project and then navigate to your file.

example to play with. We're going to import an .stl file of the "Be Happy" stamp that was made in Chapter 7 and stored on the computer desktop.

After the file is imported, a dialog box appears asking you to name and describe it (see Figure 8-37a). You have the option of keeping it private or making it public. It will then save to your online account. To return to it later, open the CNC Utility, navigate to your online

account, click the My Projects tab (see Figure 8-37b), highlight the project's thumbnail, and click Open Selected Project.

The CNC Utility Interface

Once the project is saved, the workspace in Figure 8-38 appears.

(a)

(b)

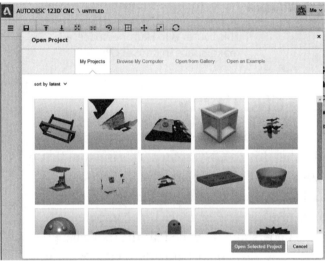

Figure 8-37 (a) The project must be named and saved before it can be worked on. (b) To return to the project later, navigate to your online 123D account, click the My Projects tab, and select its thumbnail.

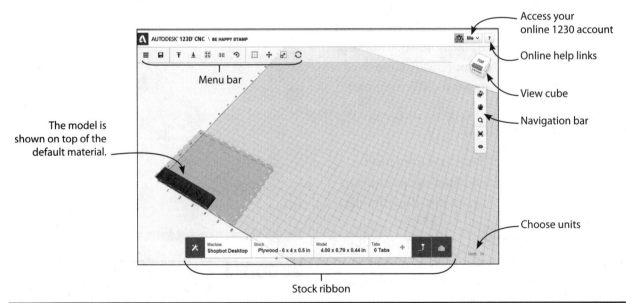

Figure 8-38 The CNC Utility interface

The interface contains a workspace on which the project is edited, a menu bar, graphics for accessing your account and online resources, a ViewCube, a navigation bar, a graphic for choosing units, and a stock (material) ribbon. They are:

- **Workspace** This is a gridded surface on which the project is edited. The model appears on a default sheet of material.

- **Menu bar** The menu bar is shown in Figure 8-39. Here, you can access the following functions:

  - *Horizontal bars* Click this icon for a submenu with the Open, Save, Save a Copy, Export Toolpath Files, Project Info, and Share Project functions (see Figure 8-40).

  - *Save* Saves the project to your online account.

- *Snap to Top* Snaps the model to the top of the stock.

- *Snap to Bottom* Snaps the model to the bottom of the stock.

- *Scale to Fit* Scales the model uniformly to fit the stock.

- *Stretch to Fit* Stretches the model along one axis to fit the stock.

- *Reset the Model* Undoes all actions at once (there is no step-by-step undo function).

- *Switch Active Plane* Toggles between the bottom plane and the current plane.

- *Precise Move* Selects one or more objects to move.

- *Precise Scale* Selects one or more objects to scale.

- *Precise Rotate* Selects one or more objects to rotate.

Figure 8-39 The menu bar

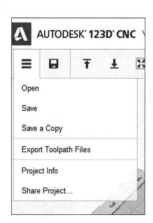

Figure 8-40 Click on the horizontal bars to access a submenu of utility functions.

Figure 8-41 Click "Me" to access your 123D. Click "?" for links to video tutorials, the support forum, blogs, and other resources.

- **Graphics for accessing your account and resources** These graphics provide links to your online account and sources of information and help (see Figure 8-41).

- **ViewCube** Click and drag the ViewCube to change its position. The model will change along with it. Click a side, such as Top or Bottom, to view the model orthographically (2D view).

- **Navigation bar** The navigation bar contains icons for moving around the workspace (see Figure 8-42).

- **Orbit** Click and drag the mouse (hold down the left button) to move around the model at any height and angle. Alternatively, orbit by holding down the right mouse button and dragging.

- **Pan** Click and drag the mouse to slide the model around the screen. Alternatively, pan by holding the mouse's scroll wheel down and dragging.

- **Magnifying glass** Click this to zoom the model around the cursor location.

- **Square** Click this to fill the screen with the model.

- **Eye** Toggle between perspective (3D) and orthographic (2D) views.

- **Unit** Click this to choose inches (in), centimeters (cm), or millimeters (mm), as shown in Figure 8-43.

- **Stock ribbon** This horizontal bar, shown in Figure 8-44, has parts for choosing the machine and stock, altering the model size, and adding tabs (pieces that hold the model to the stock while it's being carved). The Tool and Preview

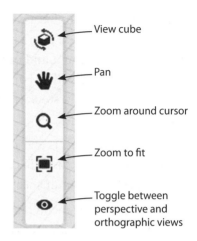

Figure 8-42 The navigation bar has icons for moving around the workspace.

Figure 8-43 Select the file's units.

icons bring up sub-ribbons. Click the wrench/screwdriver icon to return to the main ribbon. Click on this to choose the project's units.

Steps for Creating a G-Code File

Here are the steps for setting up the model to create a G-code file:

1. Choose the machine, as shown in Figure 8-45. For this example, we'll use the ShopBot Desktop, a tabletop router.

2. Choose the stock material and its dimensions, as shown in Figure 8-46. In

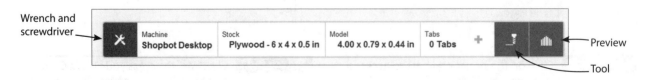

Figure 8-44 The stock ribbon contains sections for choosing options as well as icons that access sub-ribbons.

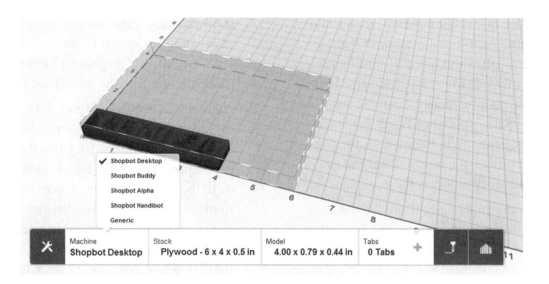

Figure 8-45 Choose the machine.

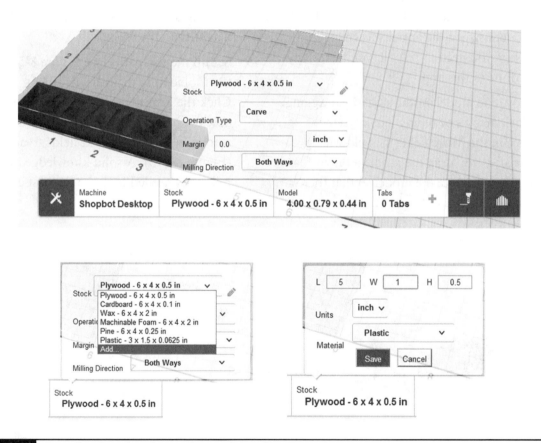

Figure 8-46 Steps for choosing the stock material and its dimensions

the Stock field, click the drop-down menu and select Add. For this example, I chose plastic for the sheet material. The stamp is 4" long by 3/4" wide by 1/2" high, so I typed dimensions a bit larger than those and then clicked Save. Now when I click the Stock section, as shown in Figure 8-47, the pop-up box reflects the new material and size, as does the stock model in the workplane.

3. Tweak the model's size if needed and adjust its position on the stock. To do this, click Model. This brings up text fields and arrowhead manipulators (see the left graphic on Figure 8-48). Change the size by dragging the manipulators or typing the new size. The model will scale proportionately along all axes unless you click the chain link; then you can scale along one axis. It's best to finalize the model's size in the app it was created in because if you need to reset it, you'll lose any sizing done in this app.

Move the model on the stock by dragging the black arrows or typing a displacement distance in the Move field (the field will appear after you drag the model a bit with a manipulator). Place the model so that there's some clearance between it and the stock's edges and then snap it to the bottom of the stock (see the left graphic on Figure 8-48). Click anywhere on the workspace to exit.

4. Add tabs, as shown in Figure 8-49. These should extend from the edges of the stock to the edges of the model. Click the Tabs plus sign and move the mouse (don't drag it) to the desired location. Click the mouse to place the tab. Click another tab onto the model and press the TAB key to rotate it 90°, as needed. Adjust the tab's size by typing in the dialog box that appears when the tab is clicked into place. Highlight it in the dialog box and type in new dimensions if needed, or highlight it in the dialog box and click the trash can to delete the tab. Put tabs on all sides. Size-wise, they should have a shorter height than the model.

5. Set up the toolpaths (see Figure 8-50). Click the tool icon to make a sub-ribbon appear. Click the gear in each sub-ribbon part to access a menu, and click the drop-down arrows in the menus' text fields to access (Figure 8-51b). A solid knowledge of CNC milling is needed to make the best choices.

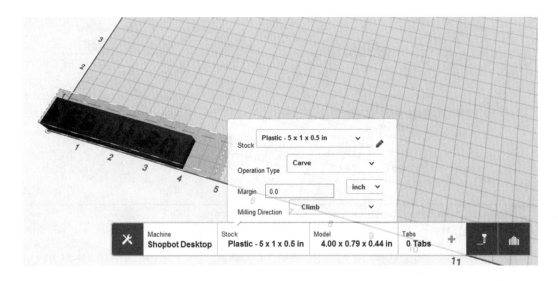

Figure 8-47   Click the Stock section to see the new entry. The stock model in the workplane reflects the new size.

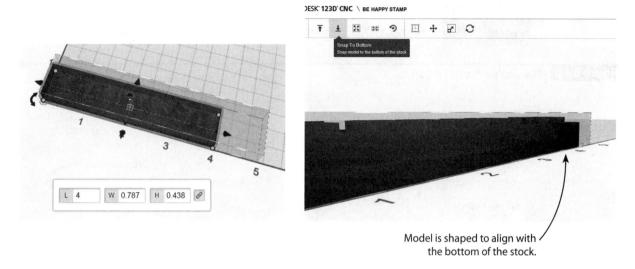

Model is shaped to align with
the bottom of the stock.

Figure 8-48 Click the Model field to change the size, if needed, and its location on the stock. Then snap it to the bottom of the stock.

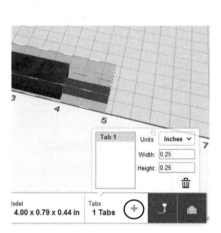

1. Click the Tabs plus sign.

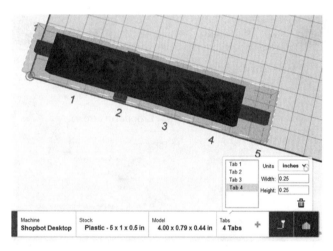

2. Move the mouse to the desired location on the model and click to place the tab. Press the TAB key to rotate it 90°, as needed.

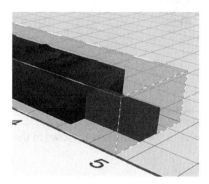

3. The tabs should have a shorter height than the model.

Figure 8-49 Adding tabs to the model.

Tool icon ———

Figure 8-50 The Toolpath sub-ribbon

Dropdown arrow

Dropdown arrow

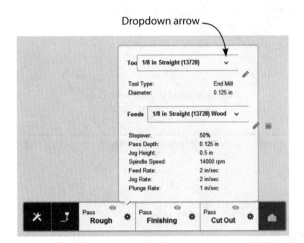

Menu choices for the Rough Pass

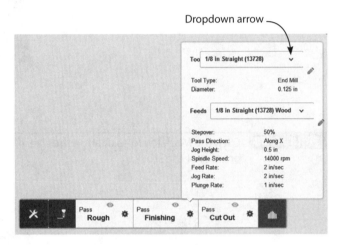

Menu choices for the Finishing Pass

Dropdown arrow

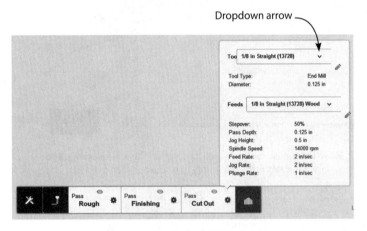

Menu choices for the Cut Out Pass

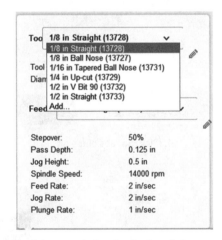

Submenu choices for the Rough Pass

Figure 8-51 Click the gear in the Rough Pass, Finishing Pass, and Cut Out Pass sections to access menu choices specific to each one. Then click the dropdown arrows in the Tools text field for submenu choices.

6. Preview the toolpath by clicking the gray eye in each of the sub-ribbon's parts. This generates a visible toolpath (see Figure 8-52). Black lines mean the tool is actively routing (cutting). Red lines mean the tool is traveling (stepping over). Study the toolpaths for errors before machining. Examples are black lines crossing through parts of the stock you don't want cut and red lines moving through unrouted stock, which will damage the bit.

7. View the toolpath animation. This can help you find errors. Click the Preview graphic on the stock ribbon to get the sub-ribbon in the top graphic of Figure 8-53, and then click Preview Settings to bring up a dialog box. Click the options you want to see, and they'll appear (bottom graphic).

Click the play button to see a simulation of the physical model getting cut (see Figure 8-54). Click the fast-forward button once to skip to the end of the rough pass, twice to skip to the end of the finish pass, and three times to skip to the end of the cut-out pass. Know that the physical model will be smoother than it looks in the animation.

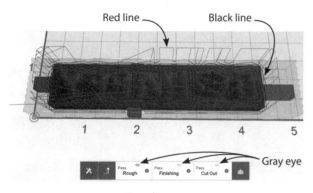

Rough pass

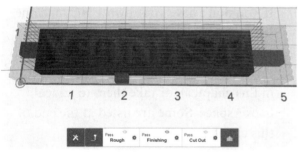

Finishing pass

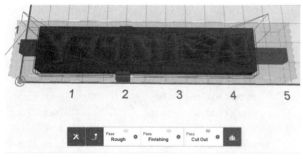

Cut-out pass

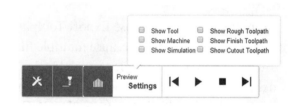

Figure 8-52 Click the gray eye in each part to generate toolpaths. Black (cutting) lines are on the letters; red (stepping) lines are above the letters. Here, toolpaths for each pass are shown.

Figure 8-53 Click the Preview graphic on the stock ribbon, click Preview Settings, and then select the options wanted. The bottom graphic shows the options checked and the picture generated.

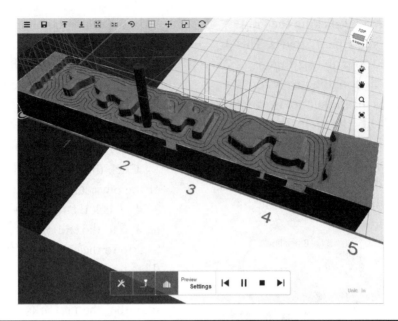

Figure 8-54 Click the play button to see a simulation of the physical model being cut.

8. From the menu bar, choose Export Toolpath Files (see Figure 8-55). Because multiple files are generated, make a folder on the local desktop for them. The whole folder needs to be copied to the PC that controls the CNC machine. This is typically done via a USB stick (thumb drive). There will be duplicates in the toolpath files. This is because .g is a generic G-code file and .sbp is a ShopBot G-code file. When setting up the CNC machine, check out the Readme.txt file for information that will need to be programmed into it (see Figure 8-56).

9. If you don't have your own CNC machine, send the files to an online service bureau via the 123D website (see the process for this in the 123D Make discussion earlier in this chapter), or take them to a local hackerspace. Some are listed at the end of this chapter.

Name ^	Date modified	Type	Size
1_Rough.g	7/1/2014 1:22 AM	G File	39 KB
1_Rough.sbp	7/1/2014 1:22 AM	SBP File	45 KB
2_Finish_x.g	7/1/2014 1:22 AM	G File	9 KB
2_Finish_x.sbp	7/1/2014 1:22 AM	SBP File	9 KB
3_Finish_y.g	7/1/2014 1:22 AM	G File	14 KB
3_Finish_y.sbp	7/1/2014 1:22 AM	SBP File	15 KB
Cutout.g	7/1/2014 1:22 AM	G File	34 KB
Cutout.sbp	7/1/2014 1:22 AM	SBP File	34 KB
Readme.txt	7/1/2014 1:22 AM	Text Document	2 KB

Figure 8-55 Choose Export Toolpath Files to generate g-code.

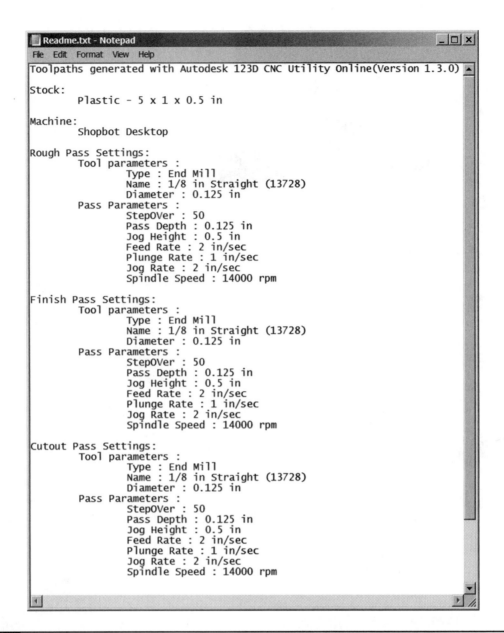

```
Readme.txt - Notepad                                        _ □ X
File  Edit  Format  View  Help
Toolpaths generated with Autodesk 123D CNC Utility Online(Version 1.3.0)

Stock:
        Plastic - 5 x 1 x 0.5 in

Machine:
        Shopbot Desktop

Rough Pass Settings:
        Tool parameters :
                Type : End Mill
                Name : 1/8 in Straight (13728)
                Diameter : 0.125 in
        Pass Parameters :
                StepOver : 50
                Pass Depth : 0.125 in
                Jog Height : 0.5 in
                Feed Rate : 2 in/sec
                Plunge Rate : 1 in/sec
                Jog Rate : 2 in/sec
                Spindle Speed : 14000 rpm

Finish Pass Settings:
        Tool parameters :
                Type : End Mill
                Name : 1/8 in Straight (13728)
                Diameter : 0.125 in
        Pass Parameters :
                StepOver : 50
                Pass Depth : 0.125 in
                Jog Height : 0.5 in
                Feed Rate : 2 in/sec
                Plunge Rate : 1 in/sec
                Jog Rate : 2 in/sec
                Spindle Speed : 14000 rpm

Cutout Pass Settings:
        Tool parameters :
                Type : End Mill
                Name : 1/8 in Straight (13728)
                Diameter : 0.125 in
        Pass Parameters :
                StepOver : 50
                Pass Depth : 0.125 in
                Jog Height : 0.5 in
                Feed Rate : 2 in/sec
                Plunge Rate : 1 in/sec
                Jog Rate : 2 in/sec
                Spindle Speed : 14000 rpm
```

Figure 8-56   The Readme.txt file contains information needed for programming the CNC machine.

Download an STL File of Your CNC Project

To get an .stl file of your CNC project, go to your online 123D account (see Figure 8-57).

Click Models, click the appropriate thumbnail, and then click the Edit/Download button. Clicking the gear icon in the thumbnail view accesses a fabrication option (see Figure 8-58).

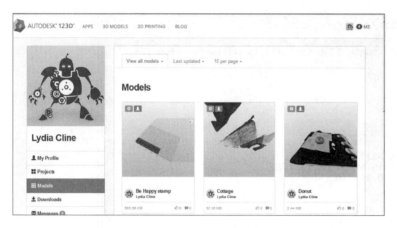

1. Click Models and then click on the appropriate thumbnail.

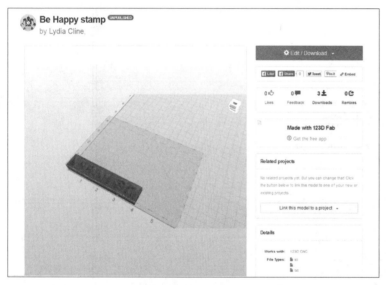

2. Click the Edit/Download button.

3. Click on "Download 3D Models."

Figure 8-57   Downloading an .stl file from the online 123D account.

Figure 8-58   Click on the thumbnail's gear to access an online fabrication option.

Summary

123D Make and the CNC Utility are powerful apps for projects made with cutting machines. You use Make to carve a model into parts for fabrication, whereas the CNC Utility generates files that a machine can read. The files can then be made locally or sent online to a third-party service bureau that can produce them for a fee.

Sites to Check Out

- **FabLab, a community hackerspace** www .fabfoundation.org/fab-labs/

- **TechShop, a membership-based DIY studio with making tools** http://techshop.ws/

- **GE's traveling garage website** www.ge.com/ garages

- **Service that matches people how have files with people how have equipment** www.100kgarages.com

- **Large custom manufacturing marketplace** www.mfg.com

- **ShopBot tools website, where you can learn about their product line** www.shopbottools. com

- **Tormach tools website, where you can learn about their product line** www.tormach.com

- **Cricut website, where you can learn about their product line and find lots of project ideas** http://us.cricut.com

- **Hackerspaces, a community-operated DIY studio with making tools** http:// hackerspaces.org

- **Make a piggy coffee table with a CNC router** www.instructables.com/id/Piggy-Coffee-Table-CNC-Router/

- **Blog entry on Cricut projects** http:// blog.123dapp.com/2014/06/123d-make-3d-diy-projects-made-with-cricut

- **Learn more about CNC** http://en.wikipedia .org/wiki/Numerical_control

- **Forum for ShopBot operators** www .talkshopbot.com/forum/index.php

CHAPTER 9

Print It! with 123D Meshmixer and MakerBot

In this chapter we'll use the Analysis tools in Meshmixer's Modify area to prepare an .stl file for printing. Then we'll hollow the file out in Meshmixer's Print area, send it to MakerBot Desktop to set preferences, and print it. We'll also look at third party printing options.

General 3D Printing Information

All printers—from consumer models that churn out plastic trinkets, to commercial ones that make metal turbines—read a digital model. You can create that model with any software, but you can't print it from a file saved in the software's native format. The file must be converted into a format the printer can read. The most common printable format is an .stl file, which is basically a shell containing a mesh model.

The .stl file and printer need software to communicate between them. The software slices the digital model into cross-sections, hence this software is also called a "slicer." Then the printer melts and deposits filament (plastic thread) in successive layers corresponding to the cross-sections. The result is a physical print with the digital model's form. This process is additive, as opposed to the subtractive process that a CNC router utilizes when cutting shapes from sheet material.

Consumer printers use a technology called *fused deposition modeling*. An extruder (also called a printer head) deposits the filament onto a build plate (also called a printer bed). The extruder is attached to a gantry, which is a system of rods and belts that moves along rails. The gantry moves the extruder in three directions: left to right along the x axis, forward and backward along the y axis, and up or down along the z axis (see Figure 9-1). A printer's resolution is the thickness of each layer, measured in microns. Thin layers provide a more detailed print than thick ones. The thinner the layer, the lower the resolution number.

Printing time can range from 20 minutes to over a day, depending on the model's size, complexity, speed setting, and printer capabilities. A 1"×1" model on MakerBot's "standard" setting prints in about 25 minutes. The amount of filament used depends on the infill setting; that is, how much material fills the model. You can set a model to print hollow, solid, or any percentage in between.

Overhangs, Supports, Rafts, and Bridges

A model needs to be supported while printed (see Figure 9-2). Overhangs, which are parts of the model that have empty space below them, present a challenge. Additive printing creates thin vertical supports, or layers of filament, to

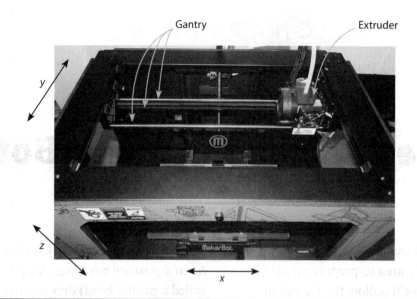

Figure 9-1 The extruder and gantry on a MakerBot Replicator 2. The extruder moves along the x, y, and z axes to deposit filament.

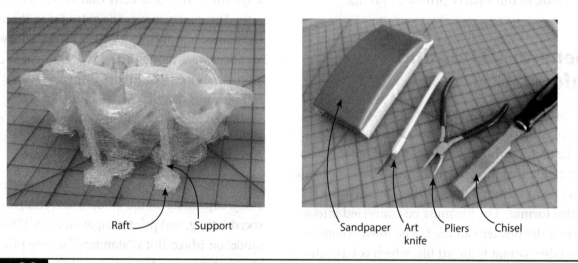

Figure 9-2 Supports are printed to hold up overhanging parts of the model. Here they hold up bracelet links. Tools may be needed to remove them. A pliers, craft knife, and sandpaper are helpful. A chisel can remove a stubborn print from the plate.

hold overhangs up while they're being formed. At the bottom of the support is a raft, which is a latticed peel-away layer. Bridges are horizontal supports stretched between gaps to hold up construction above the gap. Slicing software generates default supports, rafts, and bridges, but you can finesse by adding, deleting, or moving them to other locations. After the print is finished, supports are removed by snapping or cutting them off or by dissolving them.

The MakerBot

The MakerBot is a high-end consumer desktop printer that's compatible with Windows, Mac, and Linux operating systems. It is closed source, meaning information about its physical design and user interface is not shared. As of this writing, its models are fifth generation and non-self-serviceable. Early models had wood frames, but now they are steel-framed and sized from compact to extra large. You can view models currently sold at www.makerbot.com. Most MakerBots have one extruder. The experimental models have two, enabling the printing of two-color parts. All have different features, such as Wi-Fi connectivity, onboard camera, printed resolution, type of filament used, and heated or unheated build plate. A comparison chart is found at http://store.makerbot.com /compare. Most have a paper-thin 100-micron (0.0039" or 0.1 mm) resolution. This produces a smooth finish that doesn't need sanding or other post-production treatment. The Mini has a 200-micron resolution.

When setting the MakerBot up, place it on a sturdy table or desk in a well-ventilated area with no direct drafts (cool air affects the print). It generates noise, heat, and odor, so you might consider that when choosing a location. All models require some assembly, and all except the Mini require leveling the build plate, meaning it must be positioned parallel to, and the correct distance from, the extruder. The Mini is the simplest model (see Figure 9-3). Its few pieces that need assembly snap into place and its filament spool is smaller than those for other models.

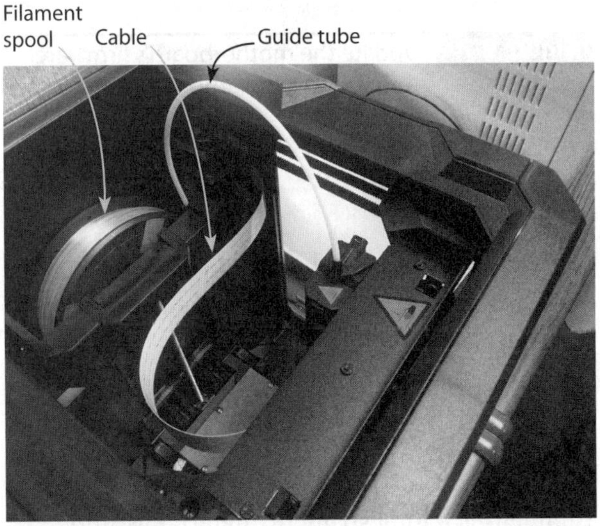

MakerBots come with a printed guide, but you may find it helpful to download it to a tablet for easy viewing during the assembly (see Figure 9-4). All guides are at www.makerbot.com/support/ guides/. My YouTube channel has two videos showing how to set up a Replicator 2.

The MakerBot plugs into a USB port on your computer, just like any other peripheral (see Figure 9-5). After 50 hours of use, lubricate the threaded rod and idler pulley by wiping them with a clean, lint-free rag and some PTFE-based grease.

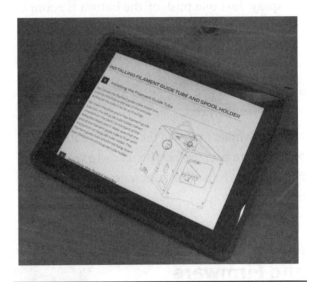

Figure 9-4 Download a guide to a tablet for easy viewing during assembly.

Filament spool Cable Guide tube

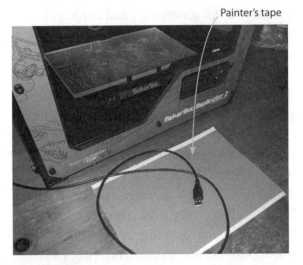

Figure 9-3 The top of a MakerBot Mini. Visible are its filament spool, guide tube, and cable.

Painter's tape

Figure 9-5 The Makerbot plugs into a computer's USB port. Here, its USB plug is resting on painter's tape used to cover the build plate.

Adherence to the Build Plate

The print must adhere to the build plate while under construction, but sometimes it adheres too much. Here are some suggestions to facilitate removal:

- Cover the build plate with painter's tape. Buy the type sold as large pieces, not strips. It must be placed smoothly onto the plate, with no wrinkles or creases.

- Spritz the build plate very lightly with hair spray. Just one push of the button is enough; more will make the plate slippery. Let it dry before printing. Clean with water and a lint-free rag.

- Use a craft spatula, putty knife, or chisel to remove the print.

Some Makers replace the acrylic plate with a glass plate because acrylic tends to warp after heavy usage, impeding adherence. You can find one on Amazon.

MakerBot Software and Firmware

The software that communicates your printing preferences is called MakerBot Desktop, and it is used for all generations and models of MakerBots. Download it at www.makerbot.com/desktop. Owners of older MakerBots should be aware that as you click through the install screens you'll see text saying the software is for fifth-generation machines. But eventually a selection window for your machine will appear (see Figure 9-6). This software is frequently updated, and notices are sent via e-mail. Firmware, which runs the printer itself, is preinstalled on the motherboard, and is also frequently updated. A firmware update notice will appear when you open MakerBot Desktop, but you can also update it by clicking Devices |

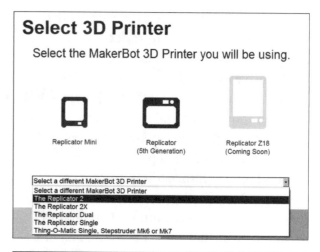

Figure 9-6 Owners of pre-fifth generation MakerBots can select their machine through this screen.

Figure 9-7 Update the motherboard's firmware through the MakerBot Desktop menu bar.

Upload Firmware on the menu bar (see Figure 9-7). The printer needs to be plugged in and turned on to do this.

Filament

Filament is a long string of material wound around a spool (see Figure 9-8). MakerBot's

Figure 9-8 MakerBot filament is a long string of plastic wound around a spool. Here, a spool is shown attached to the printer. Three unused spools are kept in their original sealed plastic for protection against humidity.

filament is made of polymer (plastic). It's nontoxic but not food safe. If you want to print beverage containers, consider printing a mold and casting a food-safe material inside it. Store filament in watertight containers with desiccant packets because humidity ruins it.

MakerBot makes four types of filament: PLA, ABS, flexible, and dissolvable, and either 1.75mm or 1.8mm wide. Some MakerBot printers use PLA; others use ABS. All are polymer and optimized for MakerBot printers, so other filament materials are not recommended.

- **ABS**   This is oil based and good for very detailed prints. It has tight tolerances and is most suited for the experimental printers. It can be used with flexible filament. ABS emanates strong fumes, so print in a well-ventilated area. It requires a heated build platform and is available in multiple colors.

- **PLA**   This is corn based, emanates mild fumes, doesn't warp as much as and is harder than ABS. It's also very shiny. PLA

is available in multiple colors, including translucent, fluorescent, glittery, and metallic. It's good for beginners because it's easy to use and performs well on most prints. It adheres to both acrylic and tape.

- **Flexible**   This is stretchy. It adheres to acrylic but not to tape. Finished prints can be modified by soaking them in hot water until they become translucent.

- **Dissolvable**   This is used with ABS filament on a heated build plate. It needs to be soaked in a limonene (chemical) bath to dissolve. Use it as a solid infill, as a soluble mold for poured materials, or for removable support structures (which results in a print with a smooth finish instead of the small scars that result from cutting off supports). It adheres well to DuPont Kapton tape, which is similar to electrical tape.

The finished product can be primed, painted, sanded, machined, glued, and drilled. This enables you to print large models in separate parts and glue or pin them together.

Export an .stl File from the 123D Apps

Design, Make, and Catch export files in their native formats directly to Meshmixer (see Figure 9-9), where they can be altered and then converted into an .stl format. Design even lets you select and send just part of the model; click the appropriate Meshmixer icon on the glyph that appears upon selection. Design also exports .stl files directly to your desktop. The CNC Utility app saves a file to your online 123D account, where you can download it for manual import into Meshmixer. Tinkercad saves .stl files to your computer.

.stl files can also be downloaded from your online 123D account and imported manually

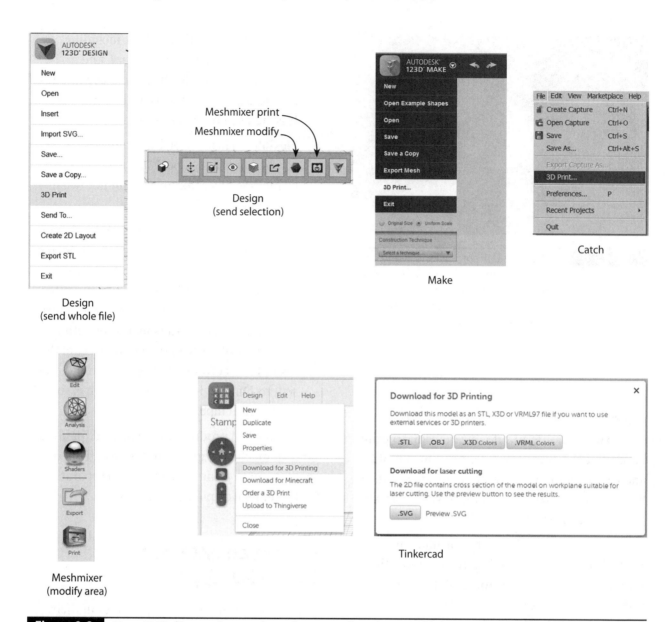

Figure 9-9 Most 123D files can be sent from their apps directly to Meshmixer. Tinkercad downloads .stl files to your desktop.

into Meshmixer. This is useful if Meshmixer crashes when a native 123D format file is sent directly to it (this sometimes happens with large Catch files). To retrieve an .stl file, click Models, click the model's thumbnail, click the Edit/Download button, and click Download 3D Models (see Figure 9-10). You'll get both an .stl and .obj file.

You Need a Printable File

Not all .stl and .obj files are printable, even if they look good in the CAD application. To be printable, a model must

- Have no holes in its mesh. This is called "watertight," meaning if you poured water into it, none would leak out. If the model has holes, the printer won't know what to fill.

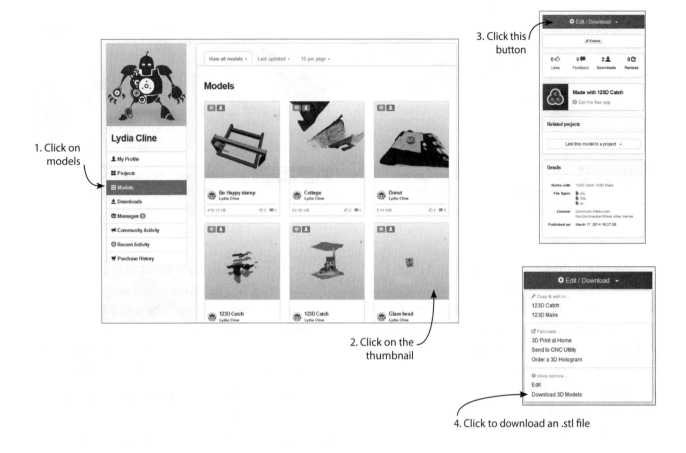

1. Click on models

3. Click this button

2. Click on the thumbnail

4. Click to download an .stl file

Figure 9-10 Files saved online can be downloaded in .stl and .obj formats.

- Be solid. The model must be one solid piece with no surface intersections. Boolean Union multiple pieces together.

- Be manifold. This means that each edge must connect to exactly two faces (a non-manifold edge connects to three or more).

- Have front-facing polygons. The normals (fronts) must face out. The faces cannot overlap or have overlapping vertices, either.

If any defects exist, the printer won't be able to read the file, and error messages will appear when you try to send it to MakerBot Desktop. It's best to address these issues in the original app. Meshmixer's analysis feature will help you fix them enough to be printable, but depending on

what and how much needs fixing, the model may change in ways you didn't expect. Preview before spending time and plastic making it.

Design Considerations

Unlike a digital CAD model that exists as pixels on a screen, a physical print has material properties and is subject to gravity. Planes have meaning to CAD software, but not to a 3D printer (see Figure 9-11).

Digital models are often scaled down before printing, but a part that is scaled down too much won't print. Features that start thin or small get thinner or smaller and must

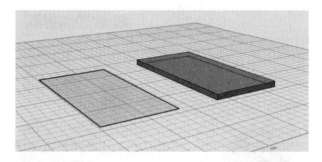

Figure 9-11 A plane can exist on a computer screen, but thickness must be added to make it printable.

Figure 9-12 Leave enough clearance between separate parts so they don't fuse together when printed.

be arbitrarily thickened or removed. Make decisions about what features to show and how to show them. This is especially important when printing with a third party, because assumptions will be made on what to change, and those assumptions may be wrong. So always be cognizant of the design's suitability for printing if a print is your end goal. Here are tips for making a printable model:

- Leave plenty of clearance (space) between moving or separate parts (see Figure 9-12). Otherwise, they'll fuse together when printed. This applies to features such as links, gears, and cogs. A clearance of 0.4mm to 0.5mm on all sides should work. Parts that must fit snugly together need between 0.1mm and 0.25mm clearance on all sides to ensure they don't easily come apart.

- Account for plastic shrinkage when making the digital model. ABS filament shrinks about 2 percent. PLA shrinks about 0.2 percent, so it's a better choice when dimensional accuracy is important. No features should be less than 2mm in size.

- Simplify complicated assemblies in the original app before printing, because fine tuning may be difficult or impossible inside Meshmixer or MakerBot Desktop. Scale the model to its final printed size in the original app, too.

- Set the digital model's units to mm or convert it to mm in the original app before printing, because that's the unit MakerBot Desktop understands. Using mm will also enable you to scale the model precisely inside that software, if you need to scale it there.

- Pointed features cannot be printed. They must be thickened.

- Thin-walled models may break during printing or shipping, or they might not print at all. This also applies to models that have a large mass connected to a thin one. .8 to 1mm generally works; 2mm is safest.

- When unsure if something can be printed, print a small, quick sample. Alternatively, send small, cheap models to a service bureau to study the results.

Meshmixer's Two Areas: Modify and Print

Meshmixer has two areas:

- **Modify**   This area enables you to significantly alter the appearance of the model, mash it up with other models, and add parts from a built-in library. Read Chapter 6 to learn more about these features. This area also contains analysis tools that study the model's suitability for printing, generate supports, and find the optimal position for construction.

- **Print** This area has options for 3D printing. Here you can also hollow out the model with one click and thicken its features. Then either send it to MakerBot Desktop to print with your own machine or upload it to a third party service bureau.

A file is sent from the Modify to the Print area by clicking the Print icon in the Modify area's vertical panel.

Import the Glass Head

 We're going to print the glass head made in Chapter 5. Choose Import from Meshmixer's launch screen to bring in an .stl file (choosing Open brings in a .mix file). Navigate to the .stl file and bring it in (see Figure 9-13). Meshmixer imports .obj, .ply, and .amf files as well. Note the vertical menu panel on the left. It contains the Analysis and Print icons we'll be using. Press and hold the spacebar to access the navigation panel (see Figure 9-14). This contains icons to zoom, pan, orbit, and fit. You can also orbit by pressing the mouse's scroll wheel down.

The model enters in the Modify area, where I made some quick changes to it. Specifically, I removed surface lumps with the Robust Smooth brush, used Plane Cut with the DelRefine Fill to slice the bottom straight, and used Extrude to make the base taller and more sturdy. A model should be as finished as possible before running the analysis tools on it.

Analyze the Glass Head

Click the Analysis icon now. It contains tools that study the model for printing suitability (see Figure 9-15). Unsuitable features need to be fixed before sending the file to MakerBot Desktop. Let's run each tool on the model.

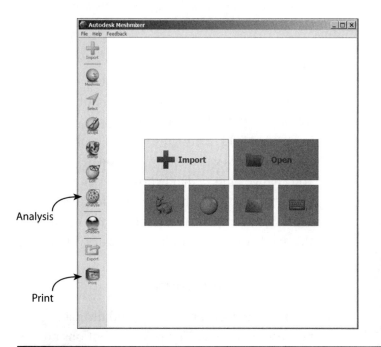
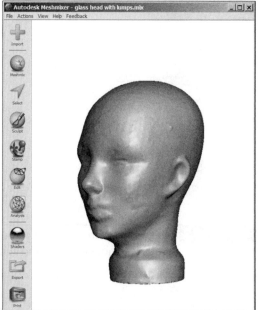

Figure 9-13 The Meshmixer launch screen and the imported glass head

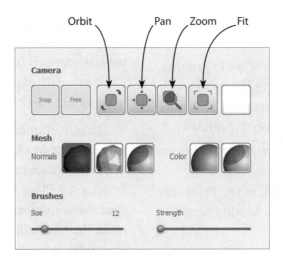

Figure 9-14   Press and hold the spacebar to access the navigation panel.

Figure 9-15   The Analysis icon contains tools that study the model for printing suitability.

Inspector

Click this to find holes in the model (see Figure 9-16a). Pins appear at the holes, and you have the option of fixing them individually or just clicking Auto Repair All. Fixing them individually gives more control over each repair; clicking Auto Repair All is faster, but may have unpredictable results. Such repairs are discussed in Chapter 6. Figure 9-16b shows a model that clearly needs a lot of fixing. No pins appear on the glass head, so click Done to exit.

Units/Scale

Click to set the model's units (they should be mm for printing). Scale the model by typing a number in one text field (see Figure 9-17). The numbers in the other fields will scale proportionately. When you change units, a dialog box appears asking if you want to keep the dimension numbers the same (for example, turn 10 into10mm) or convert the number to its equivalent in the new unit.

Measure/Dimension

Click this to measure features and distances between points (see Figure 9-18). Options are Type and Direction; a numeric distance appears after you click points.

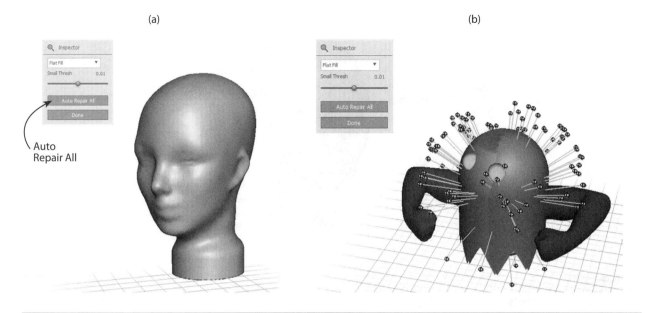

Figure 9-16 (a) Click the Inspector tool to find holes in the model. (b) Pins indicate holes.

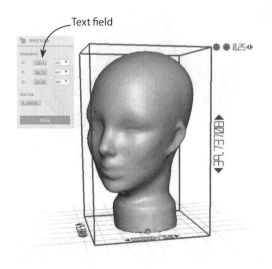

Figure 9-17 Set the models units. Change its size by typing in the text fields.

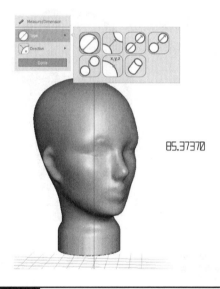

Figure 9-18 Measure distances between features.

Stability

Click this to check the surface area and volume of the model (see Figure 9-19). The ball shows the model's center of mass. A green one means it's stable and will stand up. A red one means it's unstable. You can modify it somewhat by dragging the contact tolerance slider.

Strength

Click this to see if the model is strong enough to be printed (see Figure 9-20). Solid green means it is; red areas indicate weakness. Click the Show Sections box to see more defined areas of weakness. The head is completely green, so we can click Done.

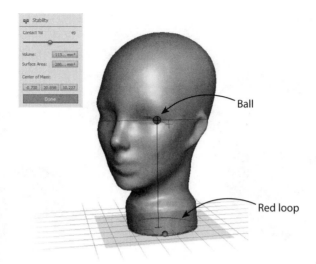

Ball

Red loop

Figure 9-19 A stability check shows whether the model will stand up when printed. The ball is the center of gravity.

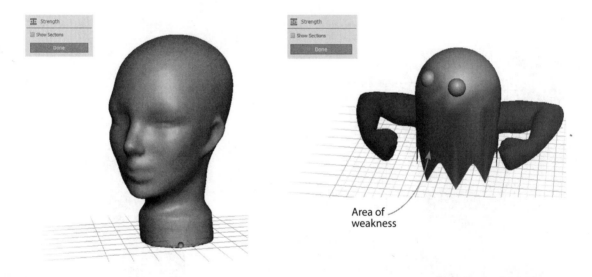

Area of weakness

Figure 9-20 A completely green model means it's strong enough to be printed. Red areas indicate weakness.

Overhangs

Click this to see overhangs that will need support during printing. They'll appear as red areas. Features that may or may not need support or may benefit from discretionary support will be outlined in blue.

Click the Optimize Orientation button at the bottom. The model will position to require the least amount of support. Then click Support All

Overhangs to generate default overhangs. These are not joined to the digital model; they are separate components.

Notice all the options in the dialog box (see Figure 9-22). To adjust them, move the sliders or click on the numbers to bring up a text field. Press the TAB key to move between fields. Changing these options changes the overhangs. You may want to increase or decrease the default overhangs, thicken them, change the angle, and

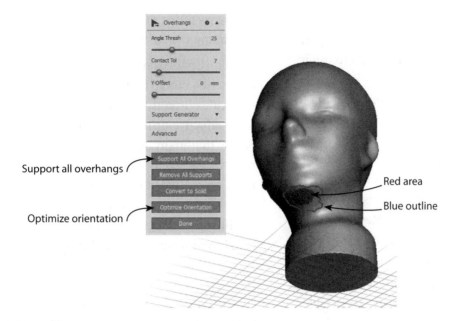

Support all overhangs

Optimize orientation

Red area

Blue outline

Figure 9-21   Red areas and blue lines show areas that need support and other overhangs.

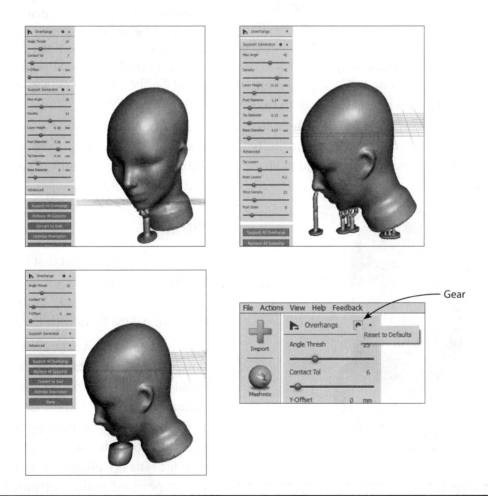

Gear

Figure 9-22   Generate and customize overhang support. Here you can see the results from different options. Click the gear to restore the defaults.

so on. Add and customize supports by clicking existing ones and dragging them to specific places on the model. If the posts tend to snap to other posts when doing this, hold the SHIFT key down and drag with the left mouse button. Clicking the same spot on a support several times generates multiple supports arrayed around it (see Figure 9-23). Cross-link supports to adjacent ones, which increases stability and serves as backup to failing supports. Remove supports by clicking them while pressing the CTRL key. Reset the defaults by clicking the gear icon at the top of the menu panel and clicking Reset to Defaults.

Good supports are critical. If they break during the printing process, everything above them will fail. When in doubt, make them thicker, add more, or connect some together by dragging and clicking. Features that are angled less than 20° from the horizontal need the most support. Generally, 3mm is a sufficient support diameter, and they should only be slightly angled (for example, no less than 60° from the horizontal). Vertical 0.35mm supports every 1cm are typically enough to span an empty space below an overhang. If there are a lot of overhangs, the print may come out "fluffy" or droopy; consider breaking the digital model

into parts for printing separately and then gluing or pinning the parts together.

Add supports after the model is scaled. If you rescale the model downstream, the supports may not scale correctly with it and end up too small to be effective or too large for easy removal. Know that many supports can be difficult to remove, especially on a delicate print. If the print breaks when you remove its supports, make a note of their settings so that you can readjust them next time. A lot of trial and error is involved in successful 3D printing.

TIP Be aware that the Optimize Orientation feature may return different results based on how the model is already oriented in Meshmixer. Try to optimally orient the model manually before applying this feature. For example, if there are features that wouldn't need support if the model's underside was flat on the build platform, then manually orient it that way instead of placing the model on its side (see Figure 9-24). The size of the build platform will also affect Optimize Orientation results.

Finally, note the Convert to Solid option. It combines multiple models together into one 3D-printable model and lets you offset distance and adjust resolution.

Slicing

This tool identifies areas that have been sliced too thin or small for the printer (see Figure 9-25). Yellow areas are too small and orange ones are too thin. Adjust them with the options box.

After analysis, save the file by clicking File | Save in the top menu bar (see Figure 9-26). Would you like to export it into a different file format? Click File | Export or the Export icon at the bottom of the menu panel and choose between .obj, .stl, .dae, .ply, .amf and .vrml. The .dae file type is Collada, an exchange format that imports into other software programs. Otherwise, click the Print icon to send the file to the Meshmixer print area.

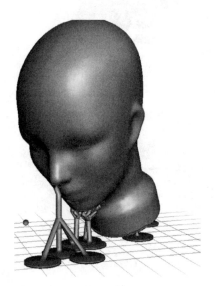

Figure 9-23 Click several times on a support to generate multiple supports arrayed around it.

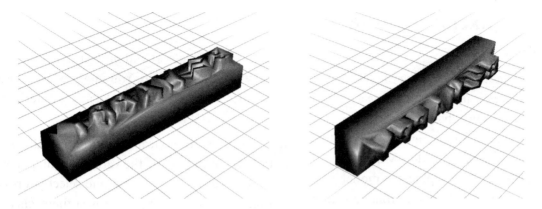

Figure 9-24   Investigate how different manual orientations affect the results of the Optimize Orientation feature. The orientation on the left won't generate supports, but the one on the right will.

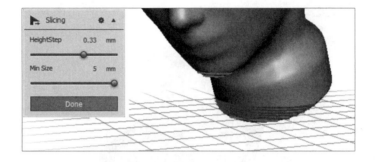

Figure 9-25   The Slicing tool identifies thin areas.

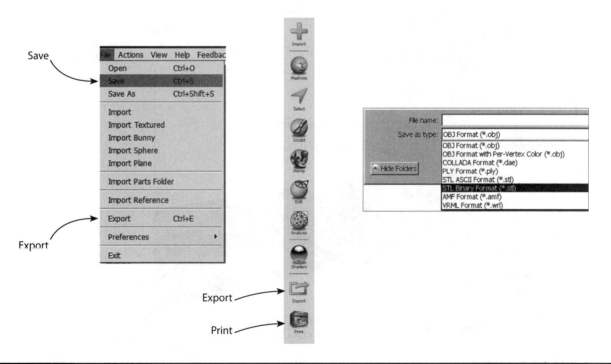

Figure 9-26   Save the file after analyzing it. Alternatively, click Export to turn it into an .obj, .stl, .dae, .ply, or .amf format. The .stl icon may look like the icon of an installed 123D app or of the MakerBot desktop icon.

The Meshmixer Print Area Interface

Clicking the Print icon brings the model and its supports to a new interface that has a menu bar at the top, a vertical menu panel, and a facsimile of the printer (see Figure 9-27). In the top graphic, the model is inside a MakerBot Mini and in the bottom graphic, the model is inside a Replicator 2. Different optimal positions and supports were also generated due to their

different manual orientations in Meshmixer. Know that a model doesn't have to be centered on the build plate. It will print just as well anywhere on the plate. You might want to print in different locations to make any painter's tape you apply last longer.

To adjust for your machine, find **Add | Edit Printers** at the top of the vertical menu panel, click the drop-down arrow, and select yours (see Figure 9-28). If your printer isn't there, click

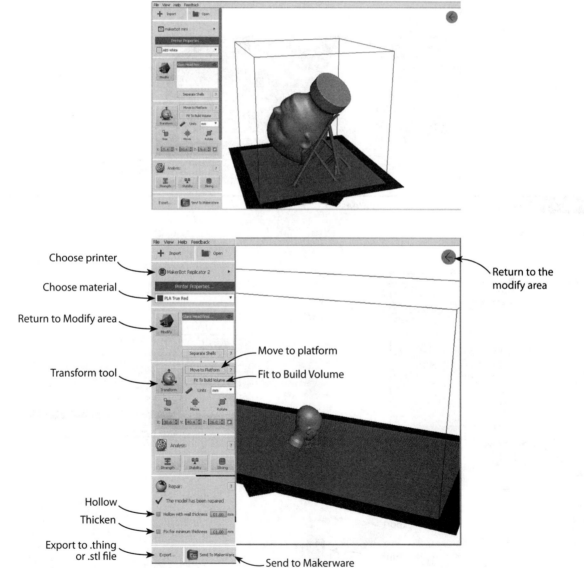

Figure 9-27 The model inside Meshmixer's Print area. At top it's in a facsimile of a MakerBot Mini; at bottom it's in a MakerBot Replicator 2

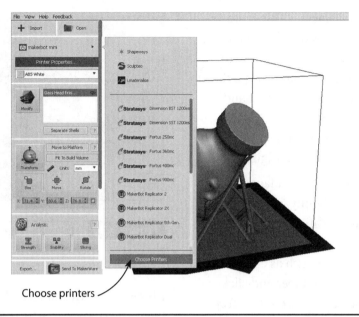

Choose printers

Figure 9-28   Scroll through the choices at Add | Printers to find yours. If it's not listed, click Choose Printers and add it there.

Choose Printers at the bottom of that window. Another window will appear. Click on the Add New Printer button at the bottom of that window and type the printer's name.

Going down the menu panel, select the printer properties (parameters and filament). Click the next field to apply color to the model, if you want. The Separate Shells option breaks the mesh into different components for models printed with multiple materials. Click Move to Platform to ensure the model sits on the build plate. A model can be printed without being moved to the plate, but the Send to Printer process will take much, much longer. Fit To Build Volume fills the printer facsimile with the model.

Meshmixer's Transform tool, shown in Figure 9-29, is accessed by clicking its icon or pressing T. It scales, moves, and rotates the model, discussed in detail in Chapter 6. However, once you do that, the supports disappear. If you clicked Fit To Build Volume, you saw that the supports disappeared, too. We'll discuss how to handle that shortly. Clicking the Modify icon or the arrow in the screen's upper-right corner returns you to Meshmixer's Modify area.

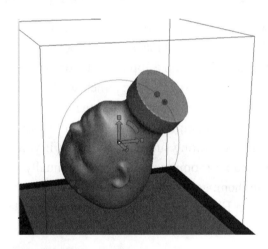

Figure 9-29   The Transform tool scales, moves, and rotates the model.

Changes made in the Print area are lost when the file is returned to the Modify area.

At the bottom of the menu panel are three features:

■ **Repair: The Model Has Been Repaired**   This is an automatic feature, not one you control. Upon import, Meshmixer analyzed every triangle and vertex in the model for problems and fixed them. A green checkmark means the repairs were successful.

- **Hollow With Wall Thickness**   This hollows out a solid model, which reduces the amount of filament used (and the cost and time to print it). Basically, it sets the infill. Type the wall thickness desired in the text field and click anywhere on the screen for it to take effect. Click Preview to see a transparent view of the hollowed-out model (see Figure 9-30).

- **Thicken Thin Parts By**   This is adaptive thickening. Meshmixer slightly widens parts of the model that are too small to print on your machine. This should only be used on models with small, pointy features such as claws, teeth, and hair. Type in a number and click somewhere on the screen for it to take effect.

Regenerate the Lost Supports

As with the Transform tool, hollowing or thickening the model makes the supports disappear. Click Add Support Structure at the bottom of the vertical menu panel (see Figure 9-31). If you don't see it, expand the window to fill the whole computer screen. It has the same tools for generating supports. However, if you have trouble zooming in enough to manually finesse them, go to File | Save Scene As, as shown in Figure 9-32, rename, and close the file.

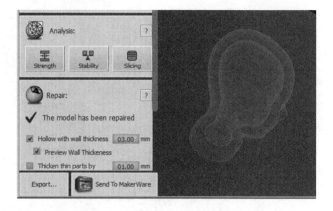

Figure 9-30   Hollow a solid model and set the wall thickness.

Reopen it in Meshmixer's Modify area. Sliding the transparency shader over it verifies that the hollowing was indeed saved (see Figure 9-33). Generate the supports again. Notice how they now go to the top of the model (see Figure 9-34).

Export the File as a THING

Click Print to return to Meshmixer's Print area. At the bottom of the vertical menu panel is an Export button. Click it for options to export the file in .stl or .thing format (see Figure 9-35). A .thing file is the MakerBot proprietary format. You can import it back into MakerBot Desktop

Figure 9-31   Click Add Support Structure to generate supports. If you don't see this option, expand the window to fill the whole computer screen.

Figure 9-32   Click Save Scene As for a new .mix file that preserves changes made in the Print area.

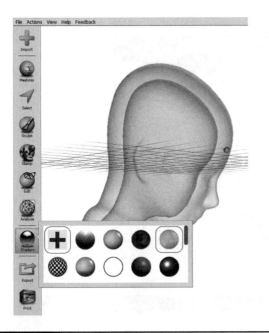

Figure 9-33 The transparency shader verifies that the model is hollow.

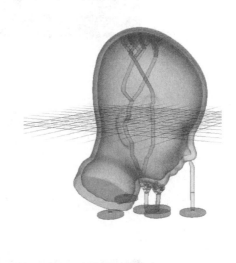

Figure 9-34 Regenerate the supports. This time they'll go to the top of the model.

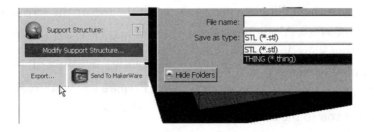

Figure 9-35 Click the Export button for options to export the file as an .stl or .thing.

later and continue to edit it. Despite the name, a .thing file is neither uploaded to nor downloaded from the Thingiverse. STL files are the dominant format there.

Now click the Send to Makerware button. It's at the bottom of the vertical menu panel, next to the Export button. The file will open inside MakerBot Desktop software (see Figure 9-36). Here, you can print it directly from the computer, export and copy it to a memory card, or tweak it some more.

The MakerBot Desktop Interface

In the upper-right corner is a sign-in icon. Use your Thingiverse log-in information. You don't have to be logged in to print, but a dialog box asking you to log in will appear and must be closed before you can proceed. Sometimes this dialog box hides; you may find it behind the Meshmixer window. You can also close out of Meshmixer once you send the model to MakerBot Desktop.

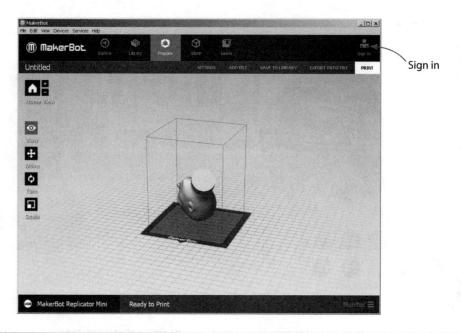

Figure 9-36   Click Send to Makerware to open the file inside MakerBot Desktop.

The model will appear on the build plate in the MakerBot Desktop interface. Along the left side are five buttons: Home View, View, Move, Turn, and Scale. Click each to access their options (see Figure 9-37):

- **Home View**   Returns you to the default view. Click the plus and minus buttons to zoom in and out.

- **Look**   Lets you view the model orthographically instead of the default perspective.

- **Move**   Change the model's position.

- **Turn**   Change the model's rotation.

- **Scale**   Change the model's dimensions.

As mentioned earlier, changing the scale may mess up the supports. But if the model imports backwards, sideways, or upside down, you can fix that with no adverse effects. If you want two heads, click the model. A gold outline indicates that it's selected. Then press CTRL-C and CTRL-V to make a copy, supports and all (see Figure 9-38).

At the top of the screen are five icons:

- **Explore** takes you to the Thingiverse.

- **Library** helps set up the printer.

- **Prepare** lets you tweak the model before printing.

- **Store** takes you to curated digital models for sale.

- **Learn** has tutorials.

If you get a "You need an internet connection" message after clicking on an icon, click the Retry button and it should find your internet connection.

When Prepare is clicked, a menu bar appears below the five icons. Here, you can choose the physical model's settings, add another file, and save the file to an online library (you need to be logged in for this). Export Print File does just that; it exports a file that can be copied onto an SD card and inserted into the MakerBot for direct printing (see Figure 9-39). The Mini is the only model without a card slot. When the export

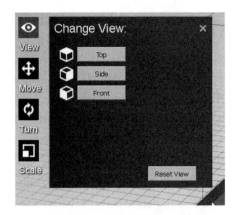

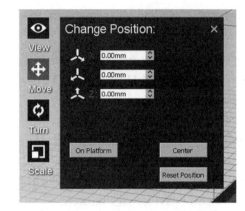

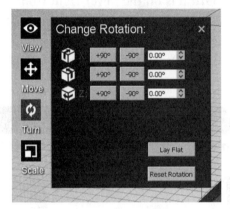

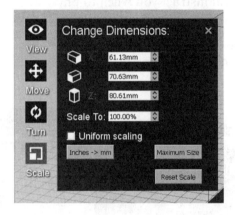

Figure 9-37 Options for changing the model's view, position, rotation, and scale.

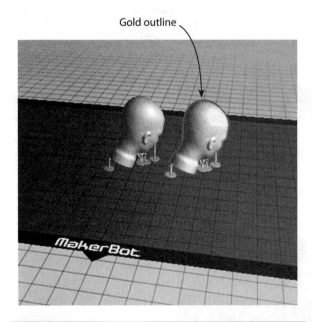

Figure 9-38 Copy the model by selecting it and pressing CTRL-C and CTRL-V.

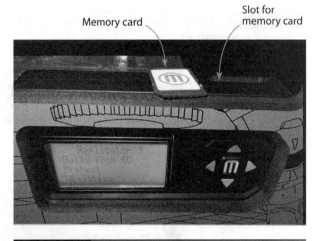

Figure 9-39 An exported file can be copied to a memory card and inserted into the MakerBot for direct printing. Here, a card is next to the slot on a fourth-generation Replicator 2. Press the left arrow to scroll to the file, and then press the M button to start printing.

is finished, a window will appear showing how much time and filament the print will take.

At the top of the screen are the File, Edit, View, Devices, Services, and Help menus. You can also set preferences and view the model orthographically. File | Examples contains premade models to print. Fifth generation machines require clicking on Devices | Change Filament each time you start a print (see Figure 9-40). The submenu has a Load Filament option that will instruct you when to physically push the filament into the guide tube. If you skip this step, the extruder will go through the motions but not deposit any filament.

At the screen's bottom-right, click on Monitor to access buttons that let you cancel, pause, and change the filament spool (see Figure 9-41). Pausing is useful for changing filament colors and almost-empty spools. Fifth generation machines pause automatically when detecting an empty spool, but earlier machines don't.

You can also pause a MakerBot directly by pressing its Action button. On a Mini this button is on the lower-right front face. Press once and it will turn from red to blue. Press again to restart. Press and hold to cancel the print. On a fourth generation Replicator 2, the

Figure 9-40   On fifth generation MakerBots, click Devices | Change Filament to load the filament each time you start a print.

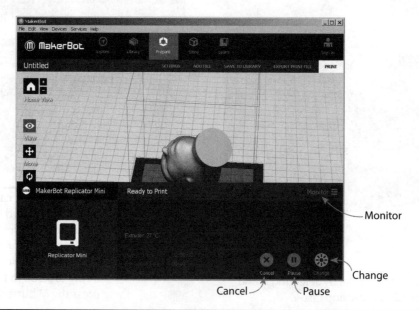

Figure 9-41   Click on Monitor at the screen's bottom-right to access buttons that let you cancel, pause and change the filament spool.

Action button is the red M. Press its left arrow, select the pause option, and press the button again to resume printing.

The Settings Menu

Click the Settings menu to choose the physical model's properties (see Figure 9-42). The resolution options are low (fast), standard, and high (slow). Models with thin supports should be done at a slow speed; models with thick supports can be printed faster. Ensure that Raft and Supports are checked if you generated them. Even if there are no supports, you might want to build a raft anyhow to help the base of a long, thin, wide model adhere better to the build plate (their corners tend to curl up when cooling). Under Advanced settings, accept the slicer defaults or type your own. You can also make and save a profile with custom settings for future prints.

Preview and Print from the Computer

Click the Print button to print from the computer. It won't start printing right away. Rather, a file will be generated and sent to it; you'll see a status bar. If the file takes a long time to send ("long time" being based on its size, but generally should take less than half an hour), check that the print is on the build platform. Recall that not being on the build platform will greatly slow down this process.

After the print has been generated, a window will appear that shows the time and filament required, print and cancel options, and a Preview Print link (see Figure 9-43). Click that link to see the toolpath (see Figure 9-44). On the left is a slider you can drag to see what the print will look like at each layer (see Figure 9-45).

Figure 9-42   Click the Settings menu to access the Print Settings. Here you choose the physical model's properties.

Figure 9-43   This window appears after the file is sent to the printer. It shows the time and filament required, has Print and Cancel buttons, and a Preview Print link.

Then click Start Print to begin printing. On the printer itself, a blinking light means the printer is warming up or waiting for user input and a solid light means the printer is working.

Finally: Ta da! The physical print (see Figure 9-46).

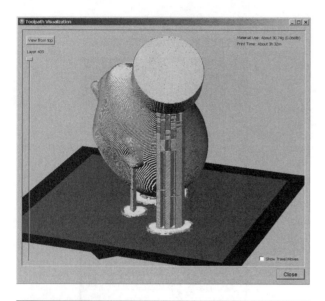

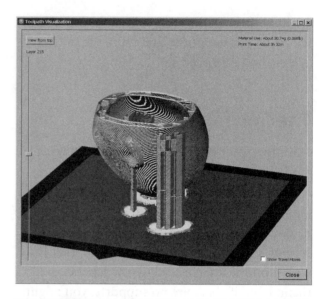

Figure 9-44 Click the Preview Print link to see the tool path.

Figure 9-45 Drag the slider to see the print at each layer.

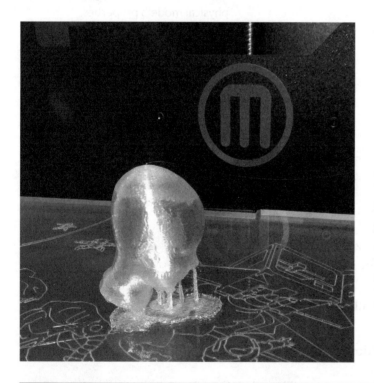

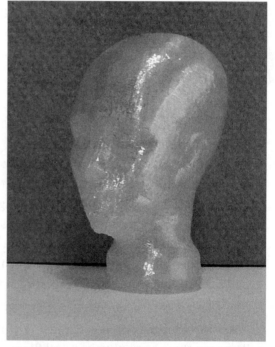

Figure 9-46 On the left the print is still on the build platform with its supports and raft. On the right, the supports have been removed and the print is standing upright.

Finding Help

A lot can go wrong during the printing process: The software doesn't connect to the printer. The extruder clogs, the printer overheats, or the build plate isn't level. Periodically check on a print that takes hours to complete to make sure all is going well. At https://www.makerbot.com /support/new/Desktop/01_MakerBot_Desktop _Knowledge_Base is a knowledge base. User forums are listed at the end of this chapter. Those who have a Makercare contract, or whose machines are still under warranty, can open a support ticket at https://www.makerbot.com /support or email support@makerbot.com.

Printing with a Service Bureau

Service bureaus are third party companies that offer online 3D printing services. They charge by material and volume. You upload your model through Meshmixer, your 123D account, or directly. You choose a material and are quoted a price. Then you enter your address and payment information. A couple of weeks later the model will arrive in the mail. Popular service bureaus are Shapeways, Sculpteo, Ponoko, and i.materialise.

Service bureaus print on expensive, commercial-grade machines, hence their prints are much higher quality than what a home printer produces. They also offer more material choices and accept more file formats. For instance, you can see all the formats Sculpteo accepts at www.sculpteo.com/en/help/#accepted -formats.

To use a service bureau, turn your file into an .stl or other format the bureau accepts. Then in Meshmixer's Print area, click the arrow in the Printer field and choose a service bureau (see Figure 9-47). The link takes you to its website. Upload your file, select a material, view the price, and check out (see Figure 9-48). The file will undergo a software analysis. If the model is not printable you'll get an error message,

Figure 9-47 Choose a service bureau.

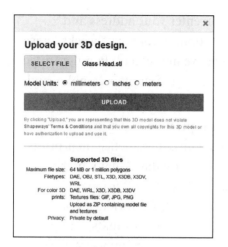

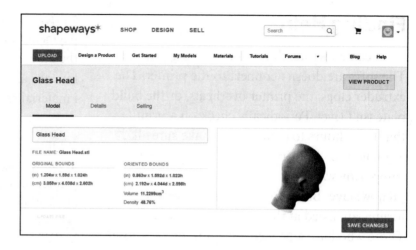

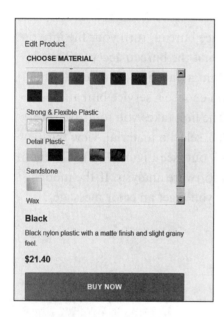

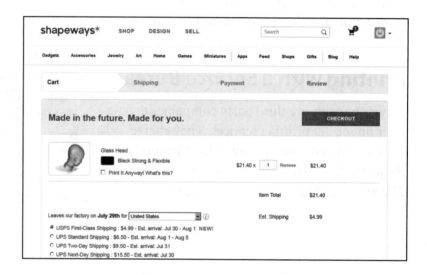

Figure 9-48 Ordering on the Shapeways website

although some bureaus may try to fix it. It will also undergo a manual review for content acceptability.

You might want to visit each of the websites directly to compare features (the URLs appear at the end of this chapter). Another option is 3D Hubs, a network of local printers. They may offer a faster turnaround or let you pick the print up yourself. Some third parties may help tweak designs as well as print them. Not all remove the supports, though; ask if that's important to you. MakerBot retail stores also have a printing service. Walk in with your .stl, .obj or .thing file on a thumbdrive.

Summary

In this chapter we discussed the 3D printing process, filament types, and what makes a model printable. We used Meshmixer to analyze a model for printability and generated supports for it. We hollowed the model out, sent it to MakerBot Desktop, set printing preferences, and then printed it on a MakerBot. We also discussed how to print through a service bureau.

Sites to Check Out

- **Video showing how to level a build plate**   www.youtube.com/watch?v=LzgcFGbMxMU

- **MakerBot operators forum**   https://groups.google.com/forum/#!forum/makerbot

- **Official Meshmixer forum**   http://meshmixer.com/forum/

- **Network of local printers**   www.3dhubs.com

- **Shapeways printing tutorials**   www.shapeways.com/tutorials?li=nav

- **Sculpteo FAQ**   www.sculpteo.com/en/help/

- **Makerbot Printshop**   www.makerbot.com/printshop. This is a tablet app for downloading and printing curated models from Makerbot's digital store.

- **Download the Autodesk Print utility**   www.apps.123dapp.com/3dprint/install.html. This app prepares files for printing, which is useful if you don't use the 123D apps or don't need Meshmixer's modifying abilities.

- **Four service bureaus websites**   www.shapeways.com, www.sculpteo.com/en/, www.ponoko.com, and www.i.materialise.com

- **Website showing prints gone bad**   http://epic3dprintingfail.tumblr.com/

- **Website that opens and converts different file types into .stl**   http://meshlab.sourceforge.net/

- **Website that edits, analyzes, scales, measures, and repairs .stl files**   http://cloud.netfabb.com/

- **Website with wall thickness suggestions for different materials and detail levels**   http://www.shapeways.com/materials/material-options

Index

Lun
mar/15- LIV
Nov. 2015

Learn by Doing!

TAB books are designed for makers, hobbyists, and experimenters.

Find out more by visiting us online at

twitter.com/TAB_DIY • tabbooks.tumblr.com • facebook.com/TAB.Books

———————

Sign up for our newsletter and get a discount when you order books
by visiting **mhprofessional.com/promo/tabbooks13**.